Benjamin Wolf, PhD

Diagnostic Techniques
for Improving Crop Production

Pre-publication
REVIEWS,
COMMENTARIES,
EVALUATIONS . . .

"**T**his is a valuable book. It covers Dr. Wolf's long-tested experiences in diagnostic techniques for soil, water, plants, maximizing pest control, the environment, and troubleshooting. To read the book in its entirety is an excellent education, particularly for growers. It can be of considerable help also to agricultural consultants, IPM personnel, and extension service workers in the ornamental, fruit, and vegetable fields. Dr. Wolf's advice is based on over 55 years of diagnosing the factors that affect marketable quality of agricultural crops. He has worked mainly with growers out of his private laboratories in New Jersey and Florida. I have worked with and near Dr. Wolf over 45 years and have seen the wide satisfaction of growers–including the best–with his services. I am happy to see Dr. Wolf disclose his valuable experiences in this book for use by others."

Norman F. Childers, PhD
Horticultural Sciences Department,
University of Florida

"In the preface the stated purpose of this book is to aid a broad range of agriculturally related personnel, ranging from grower or farm manager to professional diagnostician. The book should be required reading for college students in agriculture and a standard of reference for them after graduation.

I can confidently confirm that the author has fulfilled his goal. No available book has brought together such a wide range of practical and fundamental diagnostic material. I especially liked the inclusion of the 'on farm' tests and procedures. They bring a dimension to the book not found elsewhere. What Dr. Wolf has succeeded in doing is effectively transferring to print his vast experience as a soil and plant diagnostician. In a single volume he has compiled the material that an agriculturist needs to evaluate the soil, plant, and environmental conditions necessary for economic production and quality of a wide range of plant material.

My first impression was surprise to see Section IV (Diagnostic Procedures for Maximizing Pest Control) included in the text. After reading the chapter it was difficult to imagine it not being included. It provides the reader with material that is too often placed in a separate text or not presented with adequate depth. Very few professionals have had the personal experience with this large range of cultivated crops. None of them, to my knowledge, have been able to so effectively transfer this knowledge and experience to print."

Garth A. Cahoon, PhD
Professor Emeritus,
Horticulture Department,
Ohio Agriculture Research
and Development Center

"For the past ten years I have had the privilege of Dr. Benjamin Wolf's guidance in the production and market development programs for non-traditional crops of fruit and vegetables produced in Central and South America.

Variable production factors resulted in the loss of valuable crop potential. The explanaton of why it happened and what action was necessary to overcome the problem was limited. Dr. Wolf's input re-

Food Products Press
An Imprint of The Haworth Press, Inc.

Diagnostic Techniques for Improving Crop Production

Diagnostic Techniques for Improving Crop Production

Benjamin Wolf, PhD

Food Products Press
An Imprint of The Haworth Press, Inc.
New York • London

Published by

Food Products Press, an imprint of The Haworth Press, Inc., 10 Alice Street, Binghamton, NY 13904-1580

Library of Congress Cataloging-in-Publication Data

Wolf, Benjamin, 1913-
   Diagnostic techniques for improving crop production / Benjamin Wolf.
     p.   cm.
   Includes bibliographical references (p.   ) and index.
   ISBN 1-56022-858-X
   1. Crop improvement. 2. Crop yields. I. Title.
SB106.I47W65 1996
630'.2'087–dc20   95-31996
                                                 CIP

DEDICATION

This book is dedicated to the loving memory of my parents, who as poor immigrants to this country, took 50 acres of hardscrabble soil in southern New Jersey and by dint of hard work and all the scientific help they could get at that time, transformed the poor soil into a highly productive vegetable farm that provided a good living for us all and enough funds to provide higher education for their children and their grandchildren.

ABOUT THE AUTHOR

Benjamin Wolf, PhD, has over 55 years of research and consulting experience in soil, plant, and water analysis and crop production. Throughout his career, Dr. Wolf has made innovative analyses and discoveries. He developed practical methods of soil and plant analysis, showed that rapid soil tests can be used to evaluate fertilizer and lime requirements, used plant analyses to diagnose nutritional problems as early as 1945, used the nitrogen test as a basis for applying nitrogen fertilizers as early as the mid-1940s, introduced micronutrients into fertilizer programs in several states and foreign countries, introduced foliar feeding for cooperative growers as early as 1949, introduced the practice of fertigation to cooperative growers in 1952, and was largely responsible for the development of the cut flower business in Columbia. Most of Dr. Wolf's present consulting work is done in Guatemala, Honduras, Costa Rica, and the Dominican Republic. His research and activities have greatly increased crop yields and quality, while lifting general farming in many areas to much higher productive levels. He has published two books and over 40 articles in trade magazines and scientific journals.

CONTENTS

Foreword

Modern farming practices have evolved to levels of complexity that require knowledge over a wide breadth of disciplines for successful implementation of technical inputs. Unfortunately, there are few individual sources of reference that speak to the issues raised when trying to understand the interrelationships that exist.

Interdisciplinary research is a stated policy in most academic institutions and well it should be. We are training highly intelligent people who have the potential to do interdisciplinary work second to none. Yet, because of the degree of specialization found among today's researchers, growers often have to depend on their own trial and error experiences to develop knowledge to solve their particular problems.

In *Diagnostic Techniques for Improving Crop Production*, Dr. Wolf has presented a number of procedures across several disciplines that enable the grower to find answers involving important inputs directly in the field or through various laboratories in both the private and public sector. The book should also greatly aid the highly specialized expert to become more competent in dealing with growers' problems. By making a number of interdisciplinary techniques available, the book considerably increases the potential for finding the right answers, even if they lie outside one's specialized area.

I have had the pleasure and privilege of working with Dr. Benjamin Wolf over a period of nearly ten years. Our travels together have provided the opportunity for me to learn much about the practical aspects of plant nutrition and water management. Dr. Wolf is a great teacher in addition to being a real humanitarian. I only hope that my efforts in the information exchange have been as useful to Ben as his have been to me.

It was a direct consequence of my association with Dr. Wolf that made it possible to gain a new insight into the epidemiology of the bacterial spot disease in Florida pepper crops. Learning of such elementary concepts as the use of electrical conductivity for measuring the amount of fertilizer salt in sandy soils and the effect of soil moisture on fertilizer salt concentration enabled me to greatly expand our knowledge of this disease.

There exists a relationship between water management and outbreaks of bacterial spot disease in salt-stressed plants. The salt stress severely damages root systems and arises as a raised water table dissolves large amounts of salts from fertilizer bands placed close to the soil surface under plastic. Yet this relationship may be missed by specialists visiting fields infected with this disease since they often fail to lift a diseased plant to examine the root system. Or they fail to notice the presence of highly defined patches and gradients of infection, obviously associated with dry or poorly drained areas in the fields. By examining plant roots more carefully, and noting the striking correlation between severely damaged roots of diseased plants versus the robust root systems of healthy plants, it is possible to formulate a conclusion different from the single one which notes that the field is infected with bacterial spot, nothing more and nothing less. Also, it is by simply checking the conductivity of the soil solution in areas of different levels of infection that answers becomes available which can help explain the incidence of infection and suggest procedures to lessen its impact.

Diagnostic Techniques for Improving Crop Production provides a basis for understanding the complex relationships among physical, chemical, biological, and environmental parameters involved in plant growth that should be valuable at every level of agricultural crop production, including university research. I hope Dr. Wolf's efforts will stimulate an intensification in interdisciplinary research. The concept of Integrated Pest Management, for one, will never be fully implemented until there is an understanding at all levels of agriculture of the interrelationships manifested by modern agricultural technology.

John N. Simons, PhD
Entomologist, Virologist
President, JMS Flower Farms, Inc.
Vero Beach, FL 32690

Preface

The accumulation of scientific knowledge in the past 100 years has enabled an ever-diminishing number of growers to produce an agricultural bounty capable of supporting a population that has literally exploded. Yields that were undreamt of even 50 years ago are now commonplace.

Such production has required increased inputs of fertilizers, water, and pesticides. While the system has increased the potential for higher yields, it has made farming a much more technical enterprise—providing a greater potential for yields if inputs are used correctly, but exacting a greater cost to the grower and society if inputs are improperly used. Failure to use the various inputs properly has led to soil degradation, pollution of air and water, contaminated fruits and vegetables, greater pest damage, and undue costs of producing crops.

Fortunately, along with information on the use of increased inputs has come a wide variety of diagnostic techniques that enable the grower to provide rational use of the inputs. Rational use can largely eliminate the harmful effects while reducing the unit cost of producing a crop.

The book covers the various diagnostic techniques that can be used to provide intelligent use of the different inputs in six selections.

Section I, which consists of three chapters, covers those techniques that can impact on the intelligent use of soil. Chapter 1 in this section outlines soil testing in general with details for soil sampling. Chapter 2 details the physical characteristics that are important in crop production and outlines the tests to measure them, while Chapter 3 features the chemical aspects.

Section II outlines tests of plant materials useful for improving crop production. Chapter 4 in this section deals with the testing of seeds for vigor and purity; Chapter 5 with leaf or tissue analyses to determine plant nutrient needs; and Chapter 6 with the maturity of commodities so that they will be harvested at a time that maximizes yield and quality.

Section III deals with the application of water for crops, emphasizing those tests that determine the need for water and the suitability of water for irrigation.

Section IV summarizes diagnostic techniques that maximize pest control by counting and identifying pests, proper application of pesticides, and calibration of application equipment.

Section V outlines the methods of evaluating climate that will provide suitable conditions of light, temperature, air humidity, carbon dioxide, and freedom from air pollutants for growing, storing, and transporting plants and plant commodities.

Unlike the first five sections, which largely deal with techniques for preventing problems, Section VI provides methods for determining the causes of problems ("troubleshooting"). Properly used, approaches outlined for determining causes of poor crop performance can limit losses of the current crop and in many cases greatly improve the performance of future crops.

Background information outlining the importance of the tests is provided along with the various tests in order to help put them in perspective. Also, considerable information is provided by which a poor condition as determined by diagnostics can be changed to provide a more favorable one.

Both "on farm" and laboratory diagnostic procedures are presented. Those that can be run on the farm are covered in some detail, whereas laboratory procedures are listed. Details of laboratory procedures have been presented in a number of manuals and need not be repeated here. They are presented primarily to acquaint the grower and many of those aiding growers with suitable procedures that may be valuable for improved crop production.

To conserve space, I have used abbreviations for common scientific terms and chemical symbols for the important elements. A brief definition of terms and symbols is given as these are introduced, but in order to facilitate quick understanding at any point, all abbreviated terms and symbols are summarized in Appendix 1. Also, additional facts about the chemical elements that may be useful are presented in Appendix 2.

Common names are given for the many plants touched upon in the text, but botanical names are presented for these in Appendix 3 in order to clarify identification in some areas.

To facilitate understanding for the American reader, English and American terms are used for measurements, weights, temperatures, etc., but conversions to the metric system are given in Appendix 4. Also presented are some useful conversion facts concerning the handling of soil and water.

To help the reader get started in diagnostic techniques, I have presented sources for various materials (chemicals, kits, laboratory supplies, meters, microscopes, moisture measuring devices, irrigation controls, gauges, etc.) in Appendix 5. While this can give the reader a start, these should not be considered as recommended nor complete sources. I have not had a chance to try all the items listed and there are probably many other reliable sources.

Unfortunately, available techniques for improving crop production are inadequately used by many growers, although more diagnostic techniques are being included by the successful practitioner. The complexity of modern farming has prodded many growers to employ consultants, many of whom are more familiar with these techniques and have used them to increase crop production. As more farmers use consultants and farmers themselves become more educated in crop production and pollution control, there is a good likelihood that there will be further use of these techniques. It is the hope of the author that by bringing together many of the available diagnostic techniques, this movement will be greatly accelerated. If it does, we can expect lower unit costs for producing food and fiber, while helping to minimize pollution and at the same time favoring the conservation of soil, fertilizer, and water.

The book is directed toward growers and farm managers and those people (consultants, IPM personnel, Extension Service workers and tradespeople) who help growers achieve their potential of maximum economic yields (MEY). It also should be highly useful for such students who hope to farm or enter services that help farmers grow better crops.

Benjamin Wolf, PhD
Fort Lauderdale, Florida

Acknowledgments

The author wishes to acknowledge the help of the following: Dr. R. S. Cox, Consultant, Tallahassee, FL and Dr. J. N. Simons, Virologist/Entomologist, JMS Flower Farms, Vero Beach, FL, who were of invaluable help in writing the section on pests; Dr. George Snyder, Soil Scientist, IFAS, EREC, Belle Glade, FL, who provided valuable background for several items in the section on soils; Dr. Tom W. Embleton, Professor of Horticultural Sciences, Emeritus, University of California, Riverside, CA for the data on citrus analyses; and the many research workers who over the years provided the backup for my consulting work and in the final analysis made this book possible.

SECTION I:
SOIL DIAGNOSTICS

In a book devoted to various diagnostic procedures for improving crop production, it is logical to start with soil diagnostics. Soil properties–both physical and chemical–have a marked effect on plant growth. Diagnoses that categorize these properties are the surest way to provide an ideal environment for producing high yields of excellent quality.

Chapter 1

General Considerations

Soil is a complex natural material, consisting of mineral and organic matter, air, water, and organisms, that is capable of supporting plant growth. Many of its properties affect crop production and most of them can be effectively measured by various diagnostic procedures. Many of these procedures are readily available and cost-effective, making it possible for growers to easily modify the soil.

Testing soils before they are planted enables the grower to modify the physical and chemical characteristics so that they are ideal for a plant's growth and development. The opportunity to alter the physical properties of the soil is largely confined to the period before the crop is planted. Fortunately, physical properties change slowly, so changes made before planting the crop will be effective for at least a crop year. Some of the chemical characteristics affecting crop yield can change rapidly, but corrective measures are often available for use during the growing season, provided that periodic evaluation is made to accurately diagnose the fertility program. The programs can be altered by fertilizer additions to the soil (with sidedressings or fertigation) and or to the plant (by foliar application).

SOIL TESTS

Most soil evaluation today is done by so-called rapid tests. Between three million and four million soil samples are analyzed annually in the U.S. for available nutrients, pH, and, in some cases, organic matter (OM). At first glance, this figure appears impressive, but it points out the relatively poor application of soil testing. About one-half of the samples tested came from lawns, gardens, and urban areas. The remaining samples coming from approximately 500 million acres of cropland represent only one sample for every 250 acres–a value much too low for growers to obtain full benefits from soil testing. Only 10 to 15 percent of U.S. cropland is tested annually. Obtaining an annual soil analysis from croplands would require at least seven to ten times as many tests as are now being run.

There are compelling reasons to collect samples–at least for some elements–more frequently than once a year. These extra samples, plus those needed to be gathered from the approximately 600 million acres of pastures and grasslands and the 700 million acres of timberlands, raise the total closer to about 75 million samples annually. By failing to take sufficient numbers of soil samples, agriculture is missing one of the better opportunities to fully realize the potential returns from fertilizer nutrients while limiting damage from fertilizers to the environment.

Types of Soil Tests

The appropriate soil test varies with the crop, stage of development, climatic conditions, soil, and soil test history. The usual comprehensive, rapid soil test analysis (consisting of pH, OM, specific conductance, and available macro- and micronutrients) is desirable before planting any perennial crop or at the beginning of each growing season. Rapid tests may also supply cation exchange capacity (CEC) and percent saturation information, both of which are useful preplant tests. Tests for levels of nutrients that may be leached out of the profile need to be repeated during the crop's growth. Nitrogen is particularly vulnerable to leaching losses, but potassium and magnesium can also be lost rapidly from sandy soils or soils with low CEC.

In addition to these routine rapid soil tests, there are a number of other tests that help primarily to define the physical status of a soil. Actually, it is difficult to separate physical from chemical characteristics since some soil components affect both. OM usually analyzed with a rapid test, affects a soil's physical properties (aeration, erodibility, compaction, water infiltration, and retention), as well as its chemical properties (availability of nitrogen [N], phosphorus [P], potassium [K], and micronutrients). Soil texture–which requires a physical test–affects several chemical properties (CEC, and the availability of P, K, and micronutrients), as well as some physical properties (moisture-holding capacity [MHC], infiltration of water, and erodibility).

With the exception of texture, very few physical characteristics are now being evaluated on a large scale by soil tests. The reasons for such poor utilization of physical tests are: (1) a lack of recognition of the impact of physical characteristics upon crop yields; (2) the need for special means of sample collection; (3) the amount of time or the cost of special equipment needed to run the tests; (4) the difficulty of altering soil physical characteristics; and (5) the slowness of response to or short duration of soil treatments. Greater recognition of the importance of physical characteristics, especially soil compaction, upon crop yields may promote increased valuation of a soil's physical properties.

Collecting the Sample

A number of physical and chemical tests can be run on a sample representative of the area, although some need to be run *in situ*. Most rapid tests and tests for soil texture are run on composite samples, but it is necessary that the sample accurately represent the area under consideration. The problem of collecting a representative sample is evident when one considers that only about 1 pound of soil is finally selected to represent approximately 20 million pounds (if one sample is collected from the plow depth of every ten acres of mineral soil). Nevertheless, suitable composite soil samples can be collected by paying attention to several conditions outlined below.

Sampling Tools

Samples can be collected with trowels, spades, probes, or augers. The Hofer soil tube is useful in light soils free of gravel. It can also be used effectively in heavier soils if they are relatively dry. Mechanical samplers that can be mounted on a truck or tractor are useful for rapid multiple sampling. Typical sampling equipment is depicted in Figures 1.1, 1.2, and 1.3. Some manufacturers and distributors of soil sampling equipment are included in Appendix 5.

Selecting Areas for Sampling

The chances for collecting representative samples are measurably increased by carefully dividing the areas to be tested into homogenous units which are quite similar in previous cropping history, fertilizer-lime history, topography, soil type, or physically limiting factors. All areas that may unduly bias the results, such as fence rows, wet areas, alkaline spots, or severely eroded areas, must be omitted from the sample. If these areas are large enough to be treated separately, they need to be sampled separately and identified as potential problem areas.

A map indicating where the various samples were collected and noting special conditions should be prepared and kept for future reference. A map showing the normal division of a farm into selected sample areas is given in Figure 1.4.

Whereas as a composite sample for typical farming can represent 10-20 acres of fairly uniform fields, *Precision (site-specific) Farming* will require much more detailed and precise sampling, with subunits as small as about 2 1/2 acres. The field management philosophy attempts to maximize yields by using exact amounts of seed, fertilizer, and pesticides for particular locations by combining use of a *Global Positioning System (GPS)*, and a *Geographic Information System (GIS)*. The GPS is a satellite and receiver system that can accurately pinpoint any field, and the GIS is used for preparing digital mapping systems as well as retaining and displaying pertinent information regarding soil types, previous soil test information, and past yields. The system needs variable application equipment, capable of quickly changing rate and kind of fertilizers applied, and monitors to collect yield data corresponding to the small land units.

Having divided the farm into suitable units for sampling, the likelihood that the selected sample will represent the area in question is largely dependent on the number of probes or cores selected, the depth of these cores or probes, and how well they are mixed prior to selecting the subsample.

FIGURE 1.1. Soil sampling tubes

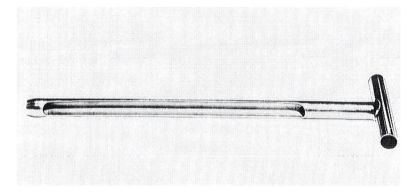

Soil sampling tubes are suitable for sampling soil largely free of gravel and with limited clay content. Model "A" is the Hofer tube with 14 1/2" profile opening and a restricted tip to supply a core slightly smaller than the tube.

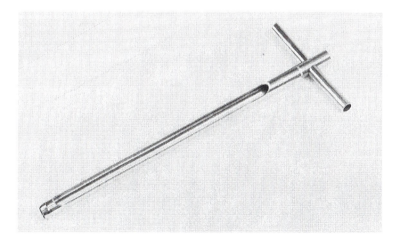

Model "B" has a 20 1/2" tube with replaceable screw tip.

FIGURE 1.1. (continued)

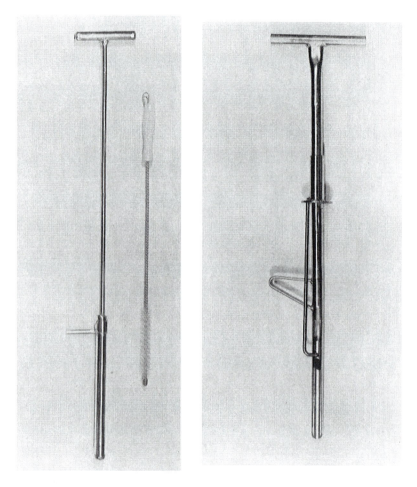

Model "C" has an 11″ tube with a 3 ft. overall length. It comes with a tube cleaning brush and a foot stand to assist penetrating hard soils.

Model "D" is similar to "C" but with a sturdier foot stand and hand plunger to eject soil sample. Photos courtesy of Nasco, P. O. Box 901, Fort Atkinson, WI 53538-0901.

FIGURE 1.2. Soil sampling augers

Soil sampling augers are better suited than tubes for sampling clay soils or those with appreciable gravel. Model "A" is the bucket type, with 3" orchard bits, 6 1/2" tubing cylinder, and 42" overall length.

FIGURE 1.2. (continued)

Model "C" is a motorized unit weighing only 8.8 lbs. and is especially suited for sampling hard or even frozen soils. Photos courtesy of Nasco, P. O. Box 901, Fort Atkinson, WI 53538-0901.

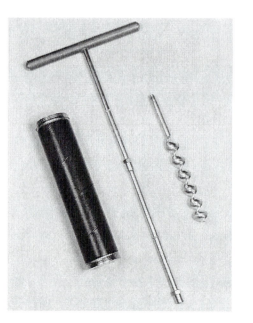

Model "B" has a 12" auger and extension rods to make an overall length of 48".

FIGURE 1.3. Cab-mounted soil sampler

Pickup- or ATV-mounted soil samplers are well suited for rapidly sampling a wide variety of soils. Up to 30 samples can be taken without leaving the vehicle, with models capable of sampling 12", 24", 36", or 48" depths. Photo courtesy of Concord Environmental Equipment, RR 1 Box 78, Hawley, MN 56549.

Number of Probes or Cores

Ten to 12 cores are usually sufficient for routine soil tests if collected from a 10-acre uniform field. Although units sampled under *Precision Farming* are often appreciably smaller than 10 acres, it is doubtful whether satisfactory composite samples can be obtained with less than about 10 cores. Greater numbers (20 or more) are needed as the number of the field's variables, such as slope, texture, OM, and fertilizer history, are increased.

Depth of Sampling

Sampling to plow depths (6-8 inches) provides most of the necessary information and is the common sample collected. If plowing

FIGURE 1.4. Map illustrating method of dividing a farm for routine sampling

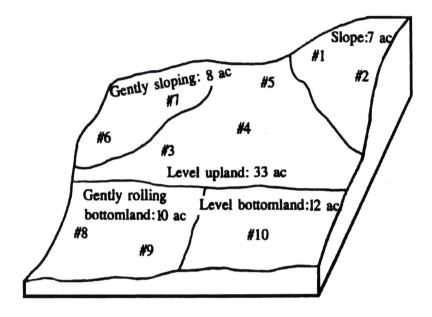

has been shallow in the past or a plow pan has developed, the grower needs to consider deeper plowing and the sample depth ought to correspond to the new program.

Subsoil samples can often be helpful for deep-rooted crops and especially if equipment is available for deep placement of lime and/or fertilizer. Deep sampling (2 feet) is also helpful in evaluating the need for N in dry climates, and for evaluating physical conditions of problem soils. Deep samples can be taken after the probe of topsoil is collected, but care must be taken to eliminate any topsoil contamination of the subsoil sample. Better evaluation of subsoil conditions can be made by digging a trench and removing soil samples from the exposed horizons.

No-till or reduced tillage farming can benefit from two samples collected from the topsoil in each area. If N has been surface-applied or applied shallowly, the samples ought to be taken at depths of 0-2 inches and 2-8 inches. If N has been injected deeper, sampling depths of 0-4 inches and 4-8 inches will be more useful.

Location of Probes

Probes can be selected at random for most sampling procedures, or several patterns (circle, closed x, or zigzag) can be used.

The above random selection is not very suitable for resampling fields that are being cropped and have had fertilizer recently banded. In such a case, a much more accurate evaluation can be made if some probes are taken directly through the band(s) and are mixed with probes from nonfertilized or broadcast-fertilized soil. A suitable pattern for such sampling requires that probes be taken about every 2 inches in the root zone on one side of the plant. The first probe is taken at the row line and others at 2 inch intervals going away from the plant and continuing to about 2 inches beyond the furthest roots. The distance between probes is modified so that all fertilizer bands are probed. All probes are taken to a depth of 6-8 inches and the process is repeated in at least three other rows.

Regardless of the sample collection method, considerable mixing of collected probes or cores is necessary before selecting the subsample to be forwarded to the laboratory. Extra mixing is necessary for samples collected from sidedressed areas, because of greater differences between probes.

Handling Samples

Handling the sample is somewhat dependent on the type of tests to be run. Generally, contamination of the sample needs to be avoided in the collection and preparation process. Common contaminants that can cause problems with several chemical tests are perspiration, cigarette ashes, and dirty pails or containers. If micronutrient tests are to be run, galvanized pails or sampling probes should be avoided, and samples should not be sieved through metal screens.

The desired method of handling and shipping samples depends to some extent on the laboratory analyzing them. Directions for handling the samples, as well as containers, shipping cartons, and directions for shipment, usually can be obtained from the testing laboratory.

All samples need to be carefully labeled with the location of the field, the crop to be planted, and the depth of sampling. All numbering and information must be on the outside of the sample bag. Additional useful information includes previous cropping and fertilizer history. If the sample is taken from an area that is already planted, information such as the age and condition of the crop, yield goals, method of fertilizer application, the kind of irrigation available, and whether the crop is to be raised in an open field, container, or bed can be of great help in outlining suitable fertility programs. This information can be supplied on separate sheets enclosed with the carton of samples or provided to the laboratory directly.

Some laboratories require samples with field moisture, especially if available N is to be determined. Such samples need to be collected when the soil is neither excessively wet or dry, placed in tightly closed moisture-proof bags, and forwarded immediately to the laboratory.

Some laboratories may require that routine samples be air dried. Drying is hastened by spreading samples on clean flat pans, plates, or wax paper in a warm, dry atmosphere. Fans can be used to facilitate the process. Avoid drying at temperatures above 100°F.

PROFILE EXAMINATION

Both physical and chemical tests are often more meaningful if the samples are collected while making an examination of the soil profile. The best time for a profile examination is when the soil is moist but not excessively wet. A preliminary examination can be made by augering and removing soil from various depths, but a much better approach is to expose the profile. This can be done by digging a trench with a shovel or backhoe. Use of a backhoe hastens the process and allows for a succession of pits to be opened as needed, thereby providing ideal moisture conditions for examination.

The trench needs to be wide enough for a person to enter and of sufficient length and depth for careful examination. Openings 2-3 feet wide and 6-8 feet long are usually satisfactory. It should be deep enough to expose the profile to the parent material. At times this may only be 2-3 feet deep, but often openings of 4-6 feet are necessary. A gradual slope or a series of steps at one end facilitates entry and exit. If the profile is more than about 3 feet deep, precautions should be taken to prevent collapse of the sides and possible entrapment.

The examination is aided by the use of a trowel or a sharp pointed tool. Samples for laboratory evaluation of both physical and chemical properties of the different horizons can be collected at this time.

Virgin Profile

Appearance will vary according to whether the soil is virgin or has been in production. (See Figure 1.5.) A virgin profile with deep A0, A1, A2, B1, and B2 layers or horizons indicates a well-developed soil with a long history. Examination of the different layers can tell much about the parent material and the climate it was subjected to if the basics of soil formation are understood.

Soil Formation

Most soils developed from rock materials during a long period of physical and chemical weathering that reduced the rocks to fragments, permitting a succession of organisms to establish themselves. Biological activity hastened and altered the development. Plants provided oxygen at deeper layers and began a cycle that moved nutrients from deeper layers to the surface in the form of litter. This litter is decom-

FIGURE 1.5. An undisturbed soil profile (A) and one modified by farming (B)

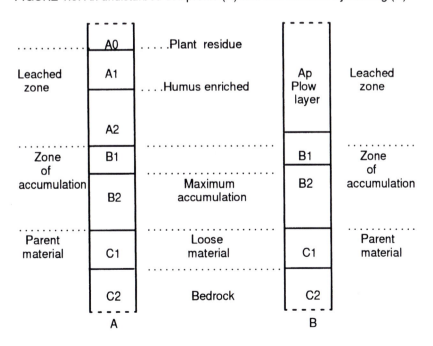

posed to humus, a relatively stable source of nutrients, water, and energy.

Climate is the dominant force in soil development, but its effects are modified by parent materials (the constituents from which the soil formed), topography, micro- and macroorganisms, and time of development. Once soil is cultivated, the usually slow forces of soil development continue, but the composition is rapidly altered by the type of cultivation, methods of soil preparation, cropping history, fertilization practices, and the amounts and quality of irrigation water applied. Because of the different forces affecting soil development and composition, great heterogeneity can exist in a soil's capability for crop production. Unlike virgin soils, where heterogeneity is often confined to large areas, the productivity of cultivated soils can vary from field to field, or even within portions of a field.

Soil Profile

The soil formation process results in layers or horizons in undisturbed soils. The appearance of these horizons, exposed as a *profile*, can be helpful in evaluating soil productivity.

The layers are formed as rainfall washes nutrients and ultrafine particles out of the top layer, or *A horizon* (zone of elution), and deposits most of them in the layer below, the *B horizon* (accumulation zone). The process is affected by the deposition of OM at the surface or in the upper portion of the soil as the remains of plants or animals accumulate. At the very top, the partially rotted residue accumulates and is referred to as the A0, or the 01 and 02 layers. Just below it is a humus-enriched layer (A1) that is dark in color as a result of humic substances moved by rainfall and organisms into the upper part of the leached zone. The lower part of the A horizon, or leached zone, is lighter in color. The maximum leached layer is referred to as the A2. There also may be a transition layer referred to as the A3, which has characteristics of the A2 and the B horizon beneath it.

The B horizon may have several layers as various components leached from the A horizon are deposited in this zone. The upper part of the B horizon (B1) will have some mineral deposition from the O and A horizons, but the layer with the greatest deposition (B2) is usually more compacted as very fine clay particles accumulate or iron concretions tend to form. The underlying material, or *C horizon* consists of parent material from which the soil is formed.

This description of a profile may not apply to some soils because of many factors. The soil may be too young, rocks too resistant, temperature too low, or rainfall too scant for the different horizons to have fully developed. Other factors limiting development are erosion that can move developed layers, and steep slopes that move water too quickly. The B horizon may be absent because of a high water table, a lack of sufficient water to move the fine particles and dissolved chemicals, or a very open parent material that allows for little settling of particles or precipitation of chemicals.

Ap or Plowed Layer

Humans, by mixing the top layers, alter much of the A horizon and sometimes some of the B horizon to that they are much more homogenous than undisturbed soil layers. The modified A horizon is referred to as the *Ap* or plowed layer. In a few soils, the depth of the O layers may be great enough so as to leave much of the A horizon intact at least for a short time span. Figure 1.5 illustrates a complete undisturbed profile and one with an Ap layer.

Examination of the Soil Profile

It is important to be observant of the following elements when examining the different profiles.

Soil Color

Some indication of the weathering and forces to which the soil has been subjected can be deduced from the color of the soil profile, even if it has been disturbed. Soils formed under predominately cold or wet conditions will tend to be dark due to the accumulation of organic matter, as with tundra or muck soils. Chernozem and chestnut brown soils also owe their dark colors to the organic matter accumulated under moderate climates that supported large grass populations (prairies) with relatively little organic matter decomposition. The poorer podzolic soils (brown forest, gray-brown, or red-yellow) developed under more humid conditions which supported forests and allowed for great depletion of bases to form acidic soils. The red latisoles developed under higher temperatures and considerable rainfall that permitted the concentration and crystallization of iron from iron-rich parent materials.

Soil drainage can be estimated from color of the soil profile. Soils that have good drainage will usually have a uniform color or a gradual change in color with increasing depth. Reddish-brown colors due to iron oxides are commonly found in highly weathered but well-aerated soils. As drainage is restricted and aeration is diminished, the normal red and brown colors tend to become paler and under extremely poor drainage conditions will be gray and blue. Soils subjected to alternate periods of excessively wet and dry conditions will have a mottled appearance.

Depth of Soil

Productivity generally is closely related to the effective depth of a soil. What might otherwise be a deep soil can be limited by the presence of impervious subsoils, poorly structured topsoils, or stratified layers. These conditions are easily revealed by close examination and careful probing at different depths.

Exactly what defines a deep soil varies with the crop, type of soil, and climatic conditions. For shallow rooted crops, an effective soil of 12-18 inches can be sufficient, especially if water can be applied as needed. For grapes and small fruits, a soil with less than about 30 inches depth would be considered shallow. Under very hot, dry conditions, a depth approaching 60 inches would be more desirable.

Soil Moisture and Moisture Distribution

A permeable soil that has been wetted a few days before the examination should be drier at the surface with the moisture gradually increasing with depth. The presence of excess moisture in certain zones may signal problems of infiltration. Dry subsoils may indicate insufficient moisture, or there may be problems with infiltration due to compaction, dense layers, or stratification.

Compact Soils

By carefully picking away at the soil with a knife or other small tool, presence of compact layers are readily distinguishable. Observations of the location and depth of the compact layer, as well as its degree of compaction, can be helpful in deciding whether ripping or subsoiling will be worthwhile.

Soil Texture

The texture of soils at various levels can be classified according to appearance and feel (see Table 1.1), or samples can be removed for laboratory evaluation.

TABLE 1.1. Textural class as indicated by physical examination

Textural Class	Characteristics as Revealed by Feel
Sand and sandy loam	Feels gritty. Single grains can be felt. Cast is formed when moist soil is squeezed but breaks easily upon handling. Dry soil crumbles. Cannot form ribbon when wet.
Sandy loam	Feels gritty but unlike sand, will dirty fingers. Compressed moist soil holds together better than sand with fine sandy loam holding better than coarse sandy loam. Other particles present but sand predominates. Cannot form ribbon when wet.
Loam	Moist soil feels smooth and not gritty. It is plastic but not sticky, and can be formed in a cast that holds with light pressure. Moist soil can be squeezed into short ribbons.
Silt loam	Moist soil is very slick and smooth but not sticky and cannot be formed into good ribbons. Has silky or soapy feeling. Makes good firm ball when compressed. Crushed dry soil can be reduced to powder.
Clay loam	Moist soil is sticky, plastic and can be worked into firm ball or ribbons about 3/4 inches long.
Clay	Very plastic and sticky when wet, but is difficult to wet when dry. Moist soil can be worked into plastic ball or into long ribbons.

Modification of data presented in University of California Leaflet # 2976, Saving Water in Landscape Irrigation. (1977).

Stratified Layers

An examination of the profile can distinguish stratified, noncompact layers that can seriously interfere with the movement of water. Such layers are usually present in sedimentary soils which have largely resulted from water movement. Abrupt changes in texture usually tend to restrict root growth, although growth is usually better as the root goes from a coarse (sandy) layer to a fine layer, rather than vice versa.

Root Systems

An examination of the profile can reveal the extent, location, and health of the root system. Such examination is especially useful in orchards and vineyards. Short, branched roots that suddenly change to a more lateral direction are indicative of compact layers, excessively wet conditions, or zones of low pH and high aluminum. Darkened, shallow roots are indicative of poor oxygen conditions typical of excessively wet soils.

Collection of Soil Samples from the Profile

Much knowledge about both physical and chemical characteristics can be gained by collecting samples from the different profile layers and analyzing them. Some of the analyses for physical properties are outlined in Chapter 2. Tests for chemical properties are discussed in Chapter 3.

Chapter 2

Physical Tests

A soil's physical properties affect: (1) amounts of water and air held; (2) infiltration and drainage; (3) erodibility of soils; (4) availability of soil nutrients, particularly nitrogen; (5) availability of water and fertilizer use; (6) penetration of roots; (7) power required to till soils; and (8) the seriousness of several soil-borne diseases.

The soil's physical characteristics are a product of the soil formation process and human handling of the soil. The characteristics related to the soil formation process may have arisen over a period of many centuries. Those influenced by humans are much more recent—some of then originating in a single growing season.

Much of the influence of physical properties upon plant growth is due to their effects on the soil's composition of air and water. To better understand the relationship between soil, air, and water, a brief description of a soil's physical composition can be useful.

PHYSICAL COMPOSITION OF THE SOIL

As illustrated in Figure 2.1, about half of the volume of a productive mineral soil is composed of solid material, with 1 to 5 percent of this being organic matter and the remainder mineral. The other half of the soil consists of pore spaces and capillaries that are filled with air and water. In an ideal situation, half of the spaces are filled with air and half with water. After rain or irrigation, all spaces may be largely filled by water. It is important that much of the water drain away rapidly and sufficient air be reintroduced. If this fails to occur, root growth and nutrient and water uptake will be affected, and the roots can become vulnerable to several root-borne diseases. On the other hand, if too little water is held, plant growth is diminished.

FIGURE 2.1. Relative amounts of the different soil components in a fertile mineral soil

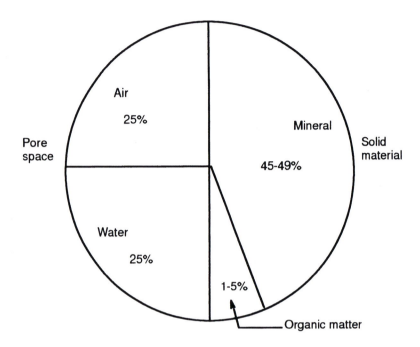

Perhaps no physical characteristics of soil are more important than those which allow for rapid exchange of air and water and still permit the soil to retain enough water for crop production. Both the exchange of air and water and the ability to retain water are greatly influenced by soil texture and soil porosity. These are not given in the usual soil texture measurements and have no bearing in classification as sands, silts, and clays.

EVALUATING PHYSICAL PROPERTIES

Soil Texture

Soil texture describes the size distribution of the soil particles, and is a reflection of the percentages by weights of the fractions of clay, silt, and sand. Clay, because of the huge surface area exposed by the numerous small particles (less than 0.002 millimeters in diameter), has the greatest effects on the basic soil properties of moisture holding (MHC), cation exchange (CEC), and buffering capacity. Silt, which is intermediate in size, has much less effect. Sand, the largest of the three particle sizes, has very little or no effect on these properties. The dimensions characterizing sand and silt vary depending on whether the International Society of Soil Science (ISSS) or the United States Department of Agriculture (USDA) scheme of classification is used. (See Table 2.1.) (Gravel and stones, defined as particles more than 2 millimeters in diameter, have little effect on basic soil properties and are excluded from the soil textural determination. They do affect soil management and use, and are noted by adding the terms "gravelly" or "stony" to textural class names.)

Besides affecting air and water exchange, soil texture influences such characteristics as (1) CEC; (2) nutrient losses by fixation, leaching, erosion, and volatilization; (3) buffering effect, which influences the maximum safe limits and ideal application methods of fertilizers, lime, and other amendments; (4) heat losses and gains; and (5) consis-

TABLE 2.1. Limits of soil particle size ranges

ISSS Scheme		USDA Scheme	
	Millimeter range		Millimeter range
Coarse sand	2.0-0.2	Very coarse sand	2.0-1.0
Fine sand	0.2-0.02	Coarse sand	1.0-0.5
Silt	0.02-0.002	Medium sand	0.5-0.25
Clay	<0.002	Fine sand	0.25-0.10
		Very fine sand	0.10-0.05
		Silt	0.05-0.002
		Clay	<0.002

tency properties, which influence such diverse characteristics as power requirements to work the soil and ease of root penetration.

The different fractions (clay, silt, and sand) vary as to exposed surface area, which greatly influences their properties. Although most soil particles are not spherical, some idea of the comparative area of the different particles can be gained from the surface area of spherical particles. Even though surface area increases tenfold as the size of spherical particles decrease tenfold, additional surface area present on the smaller particles is relatively small until particles are 0.001 millimeters or less (similar to that of clay).

Sands

Sands can be classified as very coarse (2.00-1.00 millimeters), coarse (1.00-0.50 millimeters), fine (0.50-0.25 millimeters) and very fine (0.10-0.05 millimeters). The very small sand sizes can reduce porosity, infiltration, and the amount of air held by soil. The excellent drainage and abundance of air associated with sands are properties of very coarse and coarse sands, and may be absent if the soil consists primarily of fine sands.

The importance of coarse sands for improving aeration and drainage is accentuated in container growing. The small volume of soil and the perched water table resulting from restriction of drainage by the container bottom and adhesion to the pot walls makes balancing of air and water much more difficult than in field soils. To avoid poor drainage and provide sufficient air space, particles measuring less than 0.59 millimeters (i.e., passing an NBSieve #20 [National Bureau of Standards–Sieve #20]) should be limited to about 25 percent of the volume.

The different sand sizes are determined by shaking a dry sample of soil through a nest of sieves (2 millimeters, 1 millimeter, 0.5 millimeter, 0.25 millimeter, 0.1 millimeter, and 0.05 millimeter) placed one above the other until sand particles are caught on a sieve through which they cannot pass. The different portions are collected and weighed. The determination of sand sizes is a one-time process for any given lot or area.

Clay

Not only the amount of clay but also its type has a bearing on soil properties. Clay type is determined by weathering and the rocks from which it was formed. The principal types of clay are; montmorillonite, hydrous mica, kaolinite, and the oxides and hydroxides of iron and aluminum. Many of the properties of the different clays are due to differences in their surface area.

Soils derived from montmorillonite are found in regions having a good supply of bases (K, Ca, Mg). The bases were preserved because of reduced leaching resulting from limited rain, but still capable of supporting grasslands, or insufficient drainage in the areas receiving ample rainfall.

The climate needs to be warm (semitropical or warm temperate) to ensure sufficient weathering. Montmorillonite clay retains nutrients well, but soils with considerable amounts of montmorillonite clay create problems as they are wetted and dried. When wet, they are extremely sticky, difficult to work, and have low bearing strength. The low bearing strength makes roads almost impassable, and causes building foundations to crack and fenceposts to become dislodged. As the soil dries, it becomes very hard, and huge cracks can appear.

Hydrous mica clays are found in cool climates that have enough precipitation to remove soluble salts. These clays are less subject to swelling and shrinking and provide greater bearing strength when wet than the montmorillonite clays. Although potassium is an integral part of the clay, much of it is only very slowly available as the clay weathers. In addition, this type of clay does not hold nutrients well against leaching. The kaolinite clays have only external surface areas, which, although very large (in the range of 5-10 square meters per gram of soil or 1400-2800 square feet per ounce) are small compared to the external and internal surfaces of about 800 square meters per gram (22,400 square feet per ounce) for montmorillonite clays and vermiculite.

The kaolinite clays have undergone more weathering and have a very low capacity to hold nutrients or water. They are found in regions of considerable rainfall and warm temperatures. Their ability to hold nutrients is appreciably less than that of the hydrous micas.

The oxides and hydroxides of iron and aluminum represent the ultimate in weathering. Such clays are found in tropic and semitropic soils, and are formed from parent materials rich in iron and aluminum and subject to considerable leaching. They have very poor nutrient holding capacity.

The type of clay in a given area can be identified by X ray and this determination can be highly useful, particularly when working in new areas that have had little scientific evaluation. A single determination made prior to farming new areas should be sufficient.

Textural Classification

Most soils consist of mixtures of the different sizes and are classified as sands, silts, or clays, depending on the dominant particle size. Loams are mixtures of the different-sized particles that show the properties of the different fractions in about equal proportions. Equal properties of the three different fractions are expressed even though there is less clay than silt or sand because of the greater surface area of the clay. Twelve basic soil textural classes are illustrated in Figure 2.2, a schematic diagram of the various amounts of sand (0.05-2.0 millimeters), silt (0.002-0.5 millimeters) and clay (less than 0.002 millimeters) defining the different textural classes. In addition, the adjectives very coarse, coarse, fine, and very fine are used to describe dominant sand size in the textural class of sand.

Textural classification can be helpful in choosing farm practices that maximize the efficiency of soil, fertilizer, and water use. The properties of the different textural classes influence the way soils need to be cultivated or receive fertilizer and water applications. Some of the properties affecting the cultivation and fertilization of soils are given below; those that affect fertilizer and water usage in Table 2.2; and the preferred methods of applying fertilizer and water for the different classes are given in Table 2.3.

Generally, coarse or light soils rich in sand tend to warm and drain rapidly but do not hold water or nutrients well. The fine-textured or heavy soils, rich in clay and silt, are slow to warm, but hold large amounts of water and nutrients. They tend to drain poorly and are often much more difficult to work. The medium-textured soils, with mixtures of sand, silt, and clay (loams, sandy loams, and silt loams) combine the best features of coarse and fine particles and tend to give

FIGURE 2.2. Textural triangle depicting the limits of clay, sand, and silt in the various textural classes

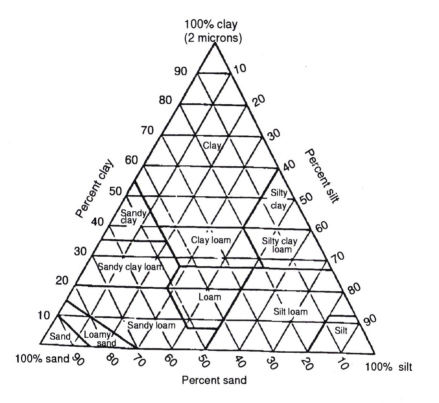

higher productivity. There are gradations in the physical characteristics of medium-textured soils, depending upon the relative concentrations of sand and clay. Those having considerably more sand (sandy loams) have a greater ability to warm up in the spring, drain more rapidly, and hold less water after rains. They are more subject to leaching, less prone to compaction and poor structure if worked wet, have lower exchange capacity, and hold fewer nutrients than those with less sand.

Determination of Textural Class

The textural class is determined from the amounts of sand, silt, and clay, in the soil (Figure 2.1). Usually, the various fractions are determined in a laboratory by the hydrometer method. First, coarse particles greater than 2 millimeters are removed by sieving. The remaining fractions are determined by dispersing them in water with a detergent and noting the rate at which the different particles settle out. Fine gravel settles out first to be followed by sand, silt, and, finally, the clay particles. The amount in suspension at various time intervals corresponds to fallout of the different particles and can be measured with a hydrometer.

Precise analysis requires prior removal of OM by oxidation with hydrogen peroxide and the removal of free calcium carbonate with acid. Treated samples have to be dried and then weighed before subjecting them to dispersion with detergent (100 g sodium hexa-meta phosphate in a solution of 1000 ml of water containing sufficient sodium carbonate to give a pH of 8.3).

A simplified procedure prepared by LaMotte Chemical Company is available through several distributors. The unit consists of three precalibrated conical tubes, dispersing and flocculating reagents, and procedures for estimating the different soil textural classes.

TABLE 2.2. Properties of several soil textural classes

Class	Bulk density	CEC	Pore space	Tilth requirements	MHC*
	%	meq/100g	%		%
Sand	1.58	<4	38	very low	9
Sandy loam	1.47	3-10	43	low	23
Loam	1.39	8-18	47	medium	34
Silt loam	1.36	10-25	49	medium	37
Clay loam	1.31	20-35	51	high	30
Clay	1.23	30-50	53	very high	45

NOTE: Average values are given for all but CEC. With the exception of bulk density, values usually will be higher as the percentage of OM increases. Bulk density tends to decrease as OM increases.

*The maximum amount of water held with free drainage.

TABLE 2.3. Preferred methods of applying water and fertilizer to different soil textural classes

Class	Fertilizer*			Water**			
	Amounts	Placed	Frequency of applic.	Amounts inches	gals	Freq.	Type of irrig.
Sands	SS LT	(1) (3)	F	0.25- 0.5	6500- 13500	VF	D,O
Loams	MS MT	(2)	I	0.5- 1.0	13500- 20000	MF	D,O,F
Clays	LS	(2)	S	0.75 1.0	20000 27000	I	D,O,F (4)

*Fertilizer
Amounts: Sands require small single applications (SS) but may need a large total (LT). Loams can tolerate a moderate single application (MS) and do best with moderate total (MT) applications. Clays can be fertilized with a single large application (LS).
Placement: (1) Avoid broadcasting fertilizers that contain N or K for sandy soils subject to leaching. (2) Place P and K in bands for loam and clay soils. (3) Use a minimal amount of N and K for pop-up fertilizer for sandy soil.
Frequency of application: F = frequent, increasing with rainfall; I = infrequent (often can be reduced to a single application if fixation is reduced by banding); S = single application for a crop, provided that the fertilizer is banded.

**Water
Amounts: Given in inches or gallons per acre to completely fill the MHC to a depth of 9".
Frequency: VF = very frequent; MF = moderately frequent; I = infrequent. Frequency should be varied as affected by climate, type of plant, and depth of rooting.
Type of irrigation: D = drip; O = overhead; F = furrow; 4 = All types can be used for clays providing delivery rate is very slow.

A quick, approximate method for determining textural classes utilizes the appearance and feel of a moist sample as it is rubbed between thumb and fingers or compressed into a ball. (See Table 1.1.)

Texture changes very slowly. Results of a determination will be relevant for years, unless the subsoil with a different texture is being brought into the plow layer by extensive erosion or by increasing plow depths.

Bulk Density

Bulk density equates the weight of soil solids to unit volume of soil or the ratio of soil mass to volume. Bulk density can be given in any terms of weight and volume but is usually expressed as grams per cubic centimeter (g/cm^3). The common bulk densities of mineral soils range between 1.0 and 1.6 g/cm^3, with the finer textured soils having the lower bulk densities because of their greater pore space. If there were no pore spaces, the bulk density of soil would be 2.65 g/cm^3 (the average density of soil mineral matter). Soils with low densities (1.0-1.4 g/cm^3) offer little impediment to the extension of roots and allow for rapid movement of air and water through the soil. Bulk densities of organic soils will be below 1.0 and that of the sphagnum moss peats will be close to 0.1 g/cm^3. Organic matter lowers bulk density because it is much lighter in weight than a similar volume of mineral matter. It also increases the aggregate stability of a soil, which tends to benefit from pore space. Compaction tends to increase bulk density. Cultivation, which tends to lower OM, also increases bulk density.

Bulk Density Determinations

Bulk density can be estimated from values given for the different soil textural classes in Table 2.2 or calculated more precisely by using Equation 1, wherein any given volume of soil can be chosen.

$$(1) \text{ Bulk density } = \frac{\text{dry weight of soil in grams (g)}}{\text{volume of soil in cubic centimeters (cc)}}$$

Laboratory procedures include selecting samples free of roots and stones by the use of core samplers. Special core samplers with a large number of brass inner rings aid in removing a core of known volume. The weight of the core is obtained and the bulk density is equal to the weight of the dried soil divided by its volume.

Another laboratory procedure suitable for soils with stones and/ or roots utilizes natural clods coated with plastic. Clods, approximately fist-sized, are coated with plastic immediately upon removal. They are immersed in a tank of water filled to an overflow outlet. The volume of displaced water, resulting from immersing the clod, equals the volume of the clod. The plastic is carefully removed from the clod before drying the soil. Again, bulk density is obtained by dividing the dry weight of the soil by its volume.

The above laboratory methods require some expertise and specialized equipment. A simplified procedure is known as the hole method, requires a minimum of tools and apparatus.

A small hole about the size of a fist is dug with a large spoon or trowel and the removed soil is carefully saved in a container for later measurement. A rubber condom or balloon is placed in the hole. Water is carefully added to the condom or balloon until it expands to conform to the contours of the hole. The water volume (cc) used equals the volume of the soil. The weight of the soil (g) removed from the hole is obtained by drying the soil in an oven for 24 hours at 220°F before weighing it.

A modification of the hole method proposed by Muller and Hamilton of the forestry department at the University of Kentucky at Lexington is more suitable for stony or hillside soils. The technique employs an expanding urethane that is obtainable from building trade stores. The expanding foam is applied from a pressurized can in a circular fashion, first filing the deepest part of the hole and continuing until the hole is filled and a slight excess is applied. Cardboard with a weight on top is placed on the surface and the foam is allowed to cure for eight hours. When the cast is removed, soil particles adhering to the surface are washed away by water and the top of the cast is trimmed flush to correspond to the top of the hole. The volume of the cast is determined by displacement of water in a large glass container. The weight of the soil is obtained as outlined under the original hole method.

In both methods, bulk density is obtained by dividing soil weight by volume as in Equation 1.

Porosity

Porosity refers to that portion of a soil not occupied by solid particles. It consists of gas- or air-filled porosity and water-filled porosity. The volume of total porosity in mineral soils ranges from 0.28 to 0.75 m3m–3, and in organic soils from 0.55 to 0.94 m3m–3. The optimum volume is probably about 0.50 m3m–3 (m3m–3 is equal to cubic meters (of pore space) per cubic meter (of soil). Ideally, about 1/2 of this volume ought to be air porosity. As a bare minimum, about 1/4 of this porosity (0.12 m3m–3) ought to be present two to three days after irrigation or heavy rainfall.

A soil's porosity is made up of different pore sizes. Rapid exchange of air and water is largely accomplished by having sufficient numbers of *macropores* (greater than 100 micrometers). Retention of water is made possible by *micropores* (less than 30 micrometers). *Mesopores* (30-100 micrometers), or intermediate-sized pores, affect both processes, but to a lesser extent.

With the exception of sand, soil particles do not tend to retain their individual identity, but combine with other particles to form various aggregates. The spaces between aggregates or sands comprise the pores or voids of the soil. These pores, largely dependent on the nature of the aggregates, are also aided by channels or tunnels left by roots, rodents, insects, earthworms, organic matter, and the alternate shrinking and swelling of the clays.

Coarse pores, or macropores, extend around the main structural soil units. The macropores need to be deep enough so that there is a rapid exchange of air and water in most of the root zone. Rain or irrigation water needs to be rapidly infiltrated into the soil, but much of it also needs to be drained rapidly so that the pores will regain sufficient air. This allows for removal of carbon dioxide and the presence of sufficient oxygen. Lack of adequate oxygen limits water and nutrient uptake and is becoming the limiting factor in many modern farm operations.

Fine pores, or micropores, extend through the interior of the soil structural units. Movement of water in the micropores is practically limited to capillary action. Micropores must be numerous enough to retain sufficient water against drainage forces and so located that they are readily reached by plant roots. Small pores not only hold water, but also allow movement of water upward through them from lower layers to help supply moisture for plants. Sufficient numbers of properly handled micropores allow for dryland farming. A soil with insufficient micropores tends to be droughty and unproductive unless there are ample rains. However, the recent use of irrigation (especially drip irrigation) has enabled efficient use of such soils.

Estimating Porosity

Pore space can be estimated from either of the following equations:

$$(2) \quad \text{\% pore space} = 100 - \left(\frac{\text{bulk density}}{\text{particle density (2.65 g/cm}^3)} \times 100 \right)$$

$$(3) \quad \text{\% pore space} = 1 - \frac{\text{bulk density}}{2.65 \text{ g/cm}^3} \times 100$$

A soil with a bulk density of 1.3 g/cm^3 would have a percent pore space of

$$\left(1 - \frac{1.3}{2.65}\right) \times 100 = 51\%$$

A more accurate determination can be made using the equation:

$$(4) \quad \text{\% pore space} = \left(1 - \frac{\text{Bulk density}}{\text{specific gravity}}\right) \times 100$$

Specific gravity can be determined in a laboratory by the pycnometer method.

Air-Filled Porosity

The air content of a soil has an important bearing on the health of roots and the uptake of nutrients and water. It has been noted that the health of conifers is greatly reduced as soil air content falls below 4 percent, that of deciduous trees and fruit trees if it falls below 10 percent, and that of nurseries and some vegetables if the air content falls below 15 percent for relatively short periods.

The air-filled porosity is of more critical concern in pot cultures because of the small volume of soil and a perched water table. While some plants and short-term growing (seed germination) can be productive in media with rather low air porosity, many pot-grown plants require media with rather high air porosities. (See Table 2.4.)

Determination of Air Porosity

Air porosity (AP) is equal to the difference between total porosity (TP) and the volume of moisture (M), which can be expressed as

(5) $AP = TP - M$ or $AP = TP - (\text{bulk density} \times M)$

$$AP = \text{percent air pore space}$$
$$TP = \text{percent total porosity}$$
$$M = \text{percent moisture}$$

Air-filled porosity can also be determined by subtracting percent field capacity (PFC) from percent pore space (P) as in the equation:

(6) $AP = P - PFC$

Percent pore space is obtained from Equations #4, #5, or #6. Field capacity is the amount of water held by a soil after gravitational water has drained away. It can be obtained by saturating a soil with water to a depth of 4 feet, allowing the water to drain while keeping the soil covered, and sampling it after 24 hours. The sample

TABLE 2.4. Desirable air-filled porosities for some ornamentals

Very high >20%	High 20-10%	Intermediate 10-5%	Low 5-2%
Azalea	African violet	Camelia	Carnation
Bromeliad	Begonia	Chrysanthemum	Conifer
Fern	Daphne	Gladiolus	Geranium
Orchid,	Foliage plants	Hydrangea	Ivy
epiphytic	Gardenia	Lily	Palm
	Gloxinia	Poinsettia	Rose
	Heather		Stocks
	Orchid,		Strelitzia
	terrestial		Turf
	Podocarpus		
	Rhododendron		
	Snapdragon		

SOURCE: White; J. W. 1973. Criteria for selection of growing media for greenhouse crops. Penn. State Agric. Expt. Journal Series #4574.

NOTE: The upper limits in each category are more suitable for smaller containers, for starting plants, repotting, and/or for plants requiring more oxygen. Mixes for starting plants or for repotting need to have more pore space because there is a normal reduction with time as organic materials decompose and all materials tend to settle.

of soil is weighed soon after collection and then again after it has been heated in an oven for 24 hours at 225°F. The PFC equals the loss in moisture divided by the dry weight of soil.

Field capacity can also be determined by subjecting a soil sample to a tension of 1/3 atmosphere in a pressure plate apparatus. This method is more rapid but more expensive because of the cost of the apparatus.

Air-Filled Porosity of Media Mixes

The air-filled porosity of media mixes or potted soils can be determined as follows:

1. The media mix is carefully placed in a 6-inch plastic pot that has the drains covered by paper towels and is repeatedly

dropped about 1 inch onto a table to settle the soil. The soil line is marked.*

2. The pot and settled soil are slowly immersed into a larger volume of water until water begins to appear at the surface. Take care to avoid allowing water to enter from the top.

3. The pot and soil are removed soon after water appears at the surface, placed in a water-tight container, and weighed in ounces or grams (W1).

4. The pot with soil is removed from the water-tight container and the soil is allowed to drain for 24 hours in a protected place free of drafts, high temperatures, and sunlight.

5. The pot and soil are then replaced in the same water-right container and reweighed (W2).

6. The soil mix and paper towels are removed from the pot, the pot cleaned, and the openings lined with thin polyethylene plastic wrap. The pot is filled to the soil line with water and reweighed (W3). The water is emptied from the pot and the pot weighed (W4).

7. The percent of air-filled pore space equals the loss in weight divided by the weight of the soil:

(7)   % air-filled pore space $-\dfrac{W1 - W2}{W3 - W4} \times 100$

Air-filled porosity, water content, volume of the soil, and the volume occupied by the soil with moisture contents above the air-dried state can be determined in a laboratory by use of an air pycnometer. The instrument employs a chamber of compressed gas at a known pressure and volume that can expand into another chamber of a larger volume. The chambers are connected with a tube containing a valve, which when opened allows the pressures in the two chambers to be equalized. According to Boyle's law the product of the pressure and volume of a gas is constant. A sample of soil is introduced and the new pressure in the expanded chamber is measured. Since the pressures of the two chambers and the volume

---

*Soil in pots with established plants can be evaluated in a similar manner.

of the second chamber are constant, the volume of the sample can be determined from the pressure of the second chamber.

Compaction

Compacting a soil can increase its bulk density and reduce the number of large pores, thereby reducing the soil's ability to provide sufficient oxygen through water infiltration. Compacted soil is also more resistant to root penetration, usually limiting the extent of a plant's root system, making it more subject to shortages of moisture and nutrients. Compacted soil also requires more draught for tillage operations.

Compaction can be present in various portions of the soil. Surface crusting results from the impact of rain or irrigation droplets as they strike the soil and the drying of a layer of dispersed particles. Topsoil compaction results from compression caused by foot traffic or machinery. Hardpans made up of densely packed sediments can exist in subsurface layers. Some can be cemented to the point of being rock-like (fragipans or ortstein). Claypans (present as subsoil layers) have a very high clay content that is resistant to penetration of air and water when wet.

Although some compaction occurs naturally (rainfall impact and subsoil compaction due to the load of topsoil), much of it results from modern farming operations. The extent of crusting is increased as the soil is deprived of cover or if the soil contains little OM to stabilize the soil aggregates. Many farm practices involving machinery, such as soil preparation, fertilizer and lime spreading, cultivation, spraying, and harvesting, can seriously compact soils. Vehicular movement, particularly on wet heavy soils primarily compresses topsoil, although subsoils may also be affected. The effect of these forces, along with slicks produced by plows and disks, often produce compact layers, or soles, that can seriously affect water drainage and restrict root penetration.

Compaction problems usually become more serious as the weight of machinery is increased, particularly if it is used on heavy soils and if the soils are wet. Fine-textured soils with little montmorillonite clay are especially subject to compaction.

Measuring Compaction

Several tools have been devised for field evaluation of compaction. The Soil Cone, Proctor and Pocket Penetrometers (available from Westgate Agronomics, Evanston IL, 60202 or Hertfodshire, HP2-7HB, England) or the soil compaction tester (available from Dickey-John, Auburn IL) give quantitative values of soil compaction. The Dickey-John apparatus is depicted in Figure 2.3.

The open-faced soil tubes, as shown in Figure 1.1 are useful in delineating problem soils. These sampling tubes are commonly used for sampling many kinds of gravel-free soil. By gradually pressing the tube into the soil, layers of compaction can be felt as more energy is required to force the tube into the soil.

FIGURE 2.3. A Dickey-John soil compaction tester

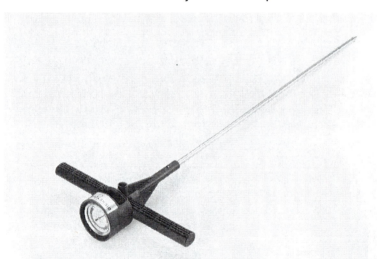

The tube is marked at 3″ intervals to a depth of 18″. Pressure required to penetrate soil to various depths can indicate whether and at what point compaction is a problem. Photo courtesy of Nasco, P. O. Box 901, Fort Atkinson, WI 53538-0901.

Results obtained by measuring compaction with any of these methods are affected by soil moisture; there is less compaction as soil moisture increases. For more meaningful results, measurements by these methods need to be made on the same soil, with varying moisture contents as reference points.

In another method of evaluating compaction, a core of soil removed by a Hofer tube is treated with a suspension of chalk. The soil must have good moisture—not too wet or too dry. With the tube turned so that the exposed soil is upward, a flat surface is cut into the entire length of the soil core with a knife. The exposed surface is then treated by adding drops of a chalk suspension (100-200 mesh limestone suspended in water) every 2 inches along the entire flat surface with a medicine dropper. Very little or no compaction is indicated if most of the solution soaks into the soil leaving only a slight deposit of chalk on the surface. Large amounts of white chalk deposit left on the surface indicate considerable compaction. Both the extent of compaction and its location is determined by this method.

Limiting Compaction

Various methods can be used to limit compaction, such as: (1) using rotations that increase OM; (2) reducing the weight of machinery; (3) limiting the use of machinery when soils are wet; (4) confining spraying and harvesting machinery to special roadways (use of long booms limits the amount of ground lost to production); and (5) spreading the weight of machinery over larger areas by using floater tires, crawler tracks, and all-wheel drive. Compaction at the surface can be largely avoided by leaving organic residues on the surface to break the force of rain or irrigation droplets. Compaction resulting from overhead irrigation can be reduced by using nozzles that produce smaller droplets.

Surface compaction can also be alleviated to a large extent by increasing the stability of surface aggregates with such farming practices as suitable rotations and the use of limestone or gypsum. Improving internal compaction problems by the use of subsoilers may be effective only in the short term unless the subsoiling is combined with the addition of gypsum and/or organic matter (OM).

Stable Aggregates

Except for sand, the soil particles seldom exist as units; usually they are combined with one another to form aggregates of different sizes and stabilities. Under natural conditions, OM is the cementing agent, but phosphoric acid, potassium and sodium silicates, and synthetic organic long-chain polymers can also bond the mineral particles into stable aggregates. There has been little practical use of materials other than natural organics for improving soil aggregate stability, although there has been a recent revival of interest in the long-chain polymers. Their ability to form very stable aggregates with very low application rates make their use practical, at least for specialized areas of high-priced crops or for specialized uses.

Increased size and stability of the aggregates enhance the soil's ability to resist crusting and erosion, and have a positive effect on water infiltration, seedling emergence, and root penetration. Soils consisting primarily of large aggregates (1-2 millimeters) tend to provide better crop growth than those with large proportions of small aggregates (less than 0.2 millimeter).

Measuring Aggregate Stability

Wet-sieving is commonly used for measuring soil aggregate stability. One procedure involves placing a sample of soil on the top of a series of sieves with openings of decreasing size and capable of measuring aggregates with diameters of 5-2 millimeters, 2-1 millimeters, 1-0.2 millimeters, and less than 2 millimeters. The sieves are slowly raised and lowered in water for a given length of time which allows for separation of the different sizes. The amounts retained on the different screens are measured by drying and weighing. The results are often expressed as percentages of the total that are larger than 0.2 millimeters.

Soil Color

Soil color can reveal some physical properties of the soil. (See Chapter 1.) Accumulation of OM, favored by cooler climates, and/or reduced soil oxygen resulting from excessive moisture or reduced tillage is responsible for much of a soil's dark color. Because of the larger amounts of OM, such soils will have a greater MHC and CEC, lower bulk densities, and a larger number of stable soil aggregates.

Deep, dark A_0, A1, or Ap layers indicate good to excellent physical properties, provided there is sufficient drainage so that oxygen will not be a limiting factor.

Podzolic sands will have a very light or bleached A1 or Ap layers due to the extreme weathering to which they have been subjected. Such soils are not only low in bases (K, Ca, Mg) but also have limited ability to hold nutrients and water. Crop performance is usually poor unless the bases are replaced and provision is made for frequent fertilization and watering. Frequent fertilization with drip irrigation allows satisfactory production from many of these soils.

Measuring Soil Color

Measurement of soil color is primarily used in soil survey work, but it can be useful in evaluating newly farmed soils by providing information about a soil's origin and some of the climatic factors influencing its development. Soil color can be preliminarily evaluated by visual examination. The measurement is enhanced by using Munsell color charts, which describe: (1) hue or dominant spectral color; (2) lightness of color; and (3) chroma, the relative strength or purity of color.

Use of a color analyzer eliminates some of the problems in matching colors. The CR-200 tristimulus color analyzer, listed as the Minolta Chroma Meter (Minolta Corp., Osaka, Japan), is capable of measuring color and noting it in Munsell notation. The instrument can be used in a laboratory or can be equipped with batteries for field operation. Data storage and a data printer increase its usefulness.

Water Infiltration

The ability of water to infiltrate soils has an important bearing on the efficiency of added water, runoff, soil erosion, and ponding. Soils in which water infiltrates too rapidly may present problems with retaining sufficient water for crop use. Those that move water too slowly can be subject to water runoff, resulting in erosion or ponding.

Infiltration is affected by conditions throughout the soil profile. At the surface, such conditions as lack of cover, soil crusts, a surface devoid of roughness, low OM, previous burning, natural resistance to wetting, and steep soil slope can adversely influence infiltration. Within the tillage layer, large amounts of clay, poor aggregation, shallow depth to impervious or slowly permeable layers, and large amounts of moisture can limit infiltration. A compact, impervious, or poorly drained subsoil also reduces water penetration. On the other hand, the presence of cracks, fissures, and old root channels improves infiltration.

Infiltration is lessened as soil particles are dispersed by rainfall or irrigation and are subsequently packed. Crop cover, mulches, and incorporated OM are effective in helping to maintain normal infiltration rates despite the effects of rainfall and irrigation. Soil compaction occurring relatively close to the surface can lead to ponded soils.

Infiltration Rate

Water infiltration varies with textural class, intensity and duration of rainfall or irrigation, soil cover, soil moisture, OM content, compaction, and the presence of certain substances in the soil which may repel water.

Soil textural class and duration of water addition are primary factors in most situations. Infiltration of sands ranges from 1-10 inches per hour, the average being closer to 2 inches. The average for heavier soils is appreciably less, at about 0.5 inches for loams and only 0.2 inches for clay soils. The infiltration rate of a sandy loam starting with a rate of about 2.5 inches per hour will slow to about 0.8 inches after one hour of infiltration. The 1.5 inches per hour starting rate of a silt loam can slow to about 0.4 inches after one hour, and the 0.8 inches per hour of a clay loam will slow to about 0.2 inches.

Measuring Infiltration

Despite the need for determining infiltration rate in order to adjust irrigation schedules, a simple accurate, rapid method is still needed. Results obtained, although capable of delineating problems, are often not directly applicable to irrigation scheduling.

A common field method employs a double ring infiltrometer that is driven into the soil. A given quantity of water is added and the time needed for the water to penetrate the soil is measured by a stop watch. Infiltration rate is given as the amount of water moving into the soil in a given unit of time, such as mm/sec. Measuring the infiltration rates of repeated applications of water can be useful in helping to evaluate infiltration over time.

The difficulty of this method and similar types that use metal plot borders and water flooding of soil is that most equipment is cumbersome, difficult to position in gravelly or stony soils, and may alter the rate of infiltration.

A simple, inexpensive infiltrometer, which overcomes some of the problems of measurement, is illustrated in Figure 2.4. For measuring the entire soil profile infiltration, drive a 3-inch diameter aluminum tube three inches into the soil and set a jug on the tube positioning the lower end of the plexiglass tube 1/2 to 3/4 inches above the soil line. Measure the amount of water infiltrating the soil in a 20-minute period.

Infiltration measurements made at increasing depths in 6 inch increments to a depth of about 48 inches can be helpful in delineating problem layers in a profile. To obtain the infiltration at various depths, auger a hole 3.5 inches in diameter to the depth required and drive the aluminum pipe in the augured hole 3 inches deeper before placing the jug. Measure the amount of water infiltrating in a 20-minute period. If it is impossible to drive the 3-inch aluminum tube in the augured hole, place the tube at the bottom of the hole and prop it up in a vertical position by placing wedges between the tube and the top of the hole. Infiltration measurement at the surface will probably also require propping up the tube.

A series of infiltrometer measurements made at increasing depths in 6-inch increments to a depth of about 48 inches can be helpful in delineating problem layers in a profile. Moisture holding capacity (MHC), affected by soil texture, porosity, and OM, will be considered in Chapter 3.

FIGURE 2.4. A simple, inexpensive infiltrometer

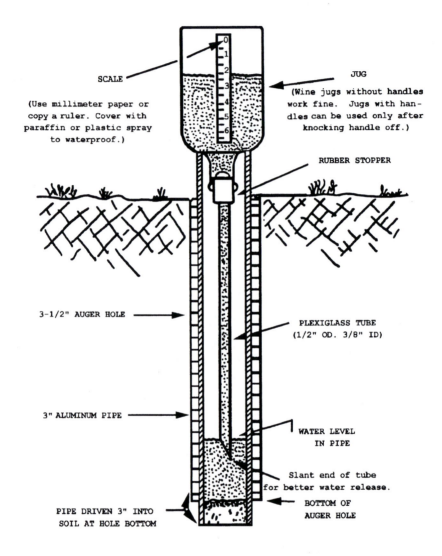

SCALE

(Use millimeter paper or
copy a ruler. Cover with
paraffin or plastic spray
to waterproof.)

JUG

(Wine jugs without **handles**
work fine. Jugs with han-
dles can be used only after
knocking handle off.)

RUBBER STOPPER

3-1/2" AUGER HOLE

PLEXIGLASS TUBE
(1/2" OD. 3/8" ID)

3" ALUMINUM PIPE

WATER LEVEL
IN PIPE

Slant end of tube
for better water release.

BOTTOM OF
AUGER HOLE

PIPE DRIVEN 3" INTO
SOIL AT HOLE BOTTOM

Source: Neja, R. A., W. E. Wildman, and L. P. Christensen. 1982. *How to Appraise Soil Physical Factors for Irrigated Vineyards.* Leaflet 2946. Div. of Agric. Sci., Univ. of California, Berkeley, CA.

AMELIORATION OF PHYSICAL PROBLEMS

Most physical problems can be prevented by adding OM (proper rotations, manures, cover crops, and sludges), using suitable land preparation methods, avoiding the use of heavy machinery or foot traffic during wet periods, using traffic lanes for machinery, using quality irrigation waters, and applying lime or gypsum. Correcting problems requires the use of OM, gypsum, lime, subsoiling, and a few synthetic polymers. The need for these corrective measures can be deduced from several of the tests already outlined above, but there are several tests which are useful in further delineating the need of both OM and gypsum.

Organic Matter

Although additions of OM provide benefits to most soils, the very heavy soils (silts and clays) and the light soils (sands) profit most from such additions. Additions to heavy soils improve porosity, bulk density, and water infiltration. The benefit to sandy soils is primarily in water retention. Additions of OM to the soil surface in the form of mulches reduce crust formation, allowing for better water infiltration and reduced erosion. Working OM into the soil improves water-stable aggregates, porosity, bulk density, and water infiltration. Placed in the subsoil behind chisels, OM can help keep soils open for longer periods.

When analysis indicates large amounts of sand or clay, high bulk density, layers of compaction, low content of stable aggregates, poor porosity, and low water infiltration, additions of OM are appropriate.

The need for adding OM can often be deduced from the values obtained in routine rapid soil analyses. More about such tests will be provided in the section Organic Matter in Chapter 3.

Growing OM in place in the form of sods, hays, or cover crops as part of long rotations is perhaps the most effective way of adding OM on a large scale. Where available, the addition of manures, sludges, and organic waste materials can also be highly effective.

Gypsum or Lime

Calcium in the forms of gypsum or lime can be expected to provide some help in correcting physical problems in many heavy soils, although they usually are of little value in improving the structure of sandy soils. Providing sufficient calcium (Ca) to saturate 70-80 percent of the exchange complex tends to group dispersed soil particles into clumps, creating better stable aggregates that provide better porosity, water infiltration, and reduced compaction.

Evaluating the Need for Gypsum to Improve Water Infiltration

The need for gypsum is suggested when heavy soils are poorly saturated with Ca (less than 65 percent), but many heavy soils can be helped by additions of gypsum if they have low amounts of stable aggregates, low porosity, poor water infiltration, or compact layers.

Usually, permeability of the soil will be greatly increased by the use of gypsum if the soil contains appreciable Na (greater than 10 percent exchangeable). The advantages decrease with the amount of Na; however, they can be appreciable in soils rich in clay but with relatively low Ca saturation (less than 65 percent), or if the addition of gypsum is combined with other procedures such as subsoiling or chiseling.

The effectiveness of adding gypsum to improve permeability can be approximated by a simple test. Two cans 3-4 inches in diameter and 4-6 inches tall are prepared by punching 1/4-inch holes in their bottoms. Circles cut from a window screen are placed directly over each hole and a circle cut from a paper towel is placed over each screen. A soil sample of at least 1 quart is collected, air dried, and then pulverized until the largest particles are no larger than coffee grounds. One heaping teaspoon of powdered gypsum is added to 1 pint of soil, mixed well and placed in a can to fill it about three fourths full. The same amount of untreated soil is similarly placed in the second can. Both cans are packed well by dropping the cans about 1 inch onto a hard surface about ten times. A circle cut from paper towels is placed over the soil in each can and then the cans are carefully filled with irrigation water. Water draining from the cans is collected and when 1/2 pint or more is collected from the treated soil, the drained amounts from the two cans are compared. If the amount from the treated soil is appreciably greater (2 times) than that from the untreated soil, there is a good possibility that the treatment will improve structure and facilitate infiltration.

Amounts of Gypsum

The approximate amounts of gypsum needed for soils containing exchangeable sodium can be calculated if the exchange capacity and percent of sodium saturation are known. For example, if the soil has an exchange capacity of 20 milliequivalents per 100 grams of soil (meq/100g), and the Na saturation is 15 percent and needs to be reduced to 5 percent, there will be 2 meq of Na that has to be replaced by gypsum [20 meq $\times$ (0.15-0.05) = 2]. The meq of gypsum is 86 and so it will take 86 $\times$ 2 or 172 milligrams (mg) of gypsum per 100 gms of soil or 0.000379 lbs per 0.22 lbs soil or about 3500 lbs of pure gypsum per plowed acre (normal furrow slice of 6 2/3 inches) weighing about 2,000,000 lbs.

The amounts needed to improve the Ca saturation can also be calculated if the CEC and percent saturation of Ca is known. Again, if the CEC is 20 milliequivalents/100 grams of soil and Ca saturation is 55 percent and a 70 percent saturation is desired, the meq of gypsum needed is 20 $\times$ (0.70-0.55) = 3. It will take 3 $\times$ 86 (milliequivalent weight of gypsum) or 258 milligrams of gypsum per 100 grams of soil, or about 2 tons per plowed acre.

Gypsum Application

Maximum benefits are obtained if the gypsum is broadcast and worked deeply into the surface soil. Some benefits can be obtained by treating a smaller portion of the total volume if treatment is confined to the volume directly under the planted row. Benefits can be obtained with about 1/4 the normal application of gypsum when applied directly under the bed or ridge in a band about 1-foot wide and worked in well before forming the bed or ridge.

Combining smaller applications of gypsum with chiseling prior to bed or ridge formation enables plants to start rapidly due to early formation of an extensive root system.

Tillage

Various types of tillage are useful in reducing compaction, surface crusting, and stratification of soils. As compaction is reduced or eliminated, the net result is better aeration and water infiltration. While beneficial for most crops, reduced compaction is absolutely essential for deep-rooted crops. Reduction of surface crusting improves water infiltration, thereby reducing erosion and ponding. Such reduction greatly benefits the emergence of crops and improves efficiency of water addition. The elimination of stratified layers in the soil improves water infiltration, increasing the efficiency of water and reducing the chances that overly wet layers will form.

Unfortunately, with the exception of eliminating stratified layers, the benefits produced by tillage may be short lived. If attempts at correction are made when soils are wet, more harm than good can result.

Special tools are needed to reduce compaction or correct stratification in deep layers. Tools used for this purpose are deep plows (moldboard, disk, and slip), rippers or subsoilers, backhoes, and trenchers.

Short-Term Effects of Deep Tillage

Often benefits obtained from deep tillage are reversed as fine particles run back together. Using tillage tools when soils are not wet increases the effective span of the treatment. Longer-lasting effects are also obtained as the deep tillage is combined with applications of OM or gypsum. Better response is obtained if the OM or gypsum is applied and worked into the soil prior to plowing or subsoiling.

Large Molecules

Improvement in soil structure can also be accomplished by treating the soil with large molecule chemicals, resulting in better water penetration, drainage, resistance to erosion, and less draught in working soils. Krilium was one of the first such materials used, but it was not readily accepted because of costs. Recently, materials that are effective with much smaller applications have appeared on the market. Compounds such as ammonium lauryl sulfate or polyacrylamide (PAM), appear to be effective with applications of ounces to a few pounds per acre. At these rates, several of these materials can easily be cost effective for many crops.

The Complete Green Company (Los Angeles, CA 90025) has been promoting the use of PAM and has obtained considerable improvement in water infiltration and seed emergence through its use. The material is sprayed onto the soil or applied through irrigation lines.

Chapter 3

Chemical Tests

Chemical evaluation of the soil can provide information as to: (1) adequacy of essential elements; (2) the presence of elements which may be harmful; and (3) factors which affect nutrient retention, release, and fixation.

Samples collected prior to planting a crop enable the grower to apply amendments and fertilizers, modifying the soil's nutritional environment so that it is ideal for the developing plant. Those collected during crop development can help insure the adequacy of nutrients while limiting wasteful excesses capable of adversely affecting potable water supplies. In cases where crop growth problems arise, chemical soil tests can often help pinpoint the cause. Tests commonly run and their importance are outlined below.

SOIL pH

The soil pH has such an important bearing on availability of nutrients, activity of microbial populations, and the efficacy of herbicides that no grower should attempt to grow a crop without first testing the pH. The apparatus for its determination is simple, rugged, and relatively inexpensive. The procedure is simple and easily conducted in the field or farm laboratory.

Ratio of Soil to Water

The pH of a soil suspension rises as the ratio of water to soil sample is increased. Common ratios of water to soil are 1:1 or 2:1, but a very narrow ratio (saturated paste) or very wide ratios of 5:1 and 10:1 are also used. A 2:1 ratio quickly wets the soil and tends to

51

give rather consistent results over relatively variable salt concentrations, and is preferred.

Determining Soil Water pH

The pH measurement, commonly run in a water suspension, can be determined by indicator dyes or pH meters. The meters are more expensive but provide more accurate results. Costs of meters will be between $50 and $1000. The $50 instrument lacks automatic temperature compensation and is not as sensitive as the more expensive instruments, but it is suitable for routine testing.

A suspension of one part soil to two parts distilled or deionized water (by volume) is prepared by stirring intermittently for one hour.

Appropriate test strips containing indicator can be dipped into the supernatant liquid of the suspension, or the suspension can be filtered and the filtrate treated with a few drops of appropriate indicator. The most commonly used indicator dye is brome cresol purple, capable of measuring pH in the range of 5.2 to 6.5+. Other indicators can be chosen to test for pH values below and above these values. Some indicator strips are capable of measuring pH over common soil pH values (4.0-8.0). (See Appendix 5 for a list of suppliers of pH dyes, strip test papers, and pH meters and buffers.)

Use of the meter requires calibration with known buffer solutions (pH 4.0 and pH 7.0) prior to measuring suspension pH. Calibration with soils of known pH can also be helpful. The electrodes are inserted into the suspension and adjusted for temperature prior to taking the reading. Stirring the suspension shortly before or during the measurement by pH meter enhances accuracy.

Suspension preparation can be eliminated for the color tests. A small amount of soil is placed in a depression of a white spot plate, the indicator solution is added dropwise until the soil is completely saturated, and the soil is stirred with a stirring rod. In a few moments, the color of the supernatant liquid at the edges of the depression can be compared with a color comparator chart.

Determining pH of a Salt Solution

The presence of salts can displace hydrogen ($H+$) and aluminum ($Al3+$) from the exchange complex which lowers pH, but this effect can be masked by the use of neutral salt solutions of 0.01 M calcium chloride or 1 N potassium chloride instead of water.

The commonly used calcium chloride solution (0.01 M $CaCl_2\cdot$ $2H_2O$) is made by dissolving 1.47 grams in 1000 milliliters of water. The soil suspension is prepared as above except that the calcium chloride solution is substituted for water. Measurement is made with a glass electrode meter, but it is necessary to use a different set of interpretative values than that used for the water pH.

Using pH Values

The preferred pH ranges for a number of crops are given in Table 3.1. In order to obtain desirable yields, the grower may use liming materials (see Table 3.2) if pH values are too low or acidifying materials (see Table 3.3) to correct high pH values.

Changing Soil pH

The amount of limestone needed to raise a low pH to a satisfactory one can be estimated from OM and textural class (see Table 3.4) or, more accurately, by buffer pH values.

Different buffer methods are chosen for determining the need for liming materials. The Adams and Maclean (AM) method is useful for soils of low CEC. The Shoemaker, McLean, and Pratt (SMP) method is used for acid soils containing appreciable aluminum, and the Woodruff buffer is used for highly buffered, strongly acid soils. Many soil testing laboratories are equipped to run the suitable buffer test and make the necessary lime recommendations.

Sulfur (S) or one of the acidifying materials listed in Table 3.3 can be applied if pH values need to be lowered. The amounts of S per acre necessary for acidification of several soil textural classes can be estimated from Table 3.5.

Ideally, the crop needs to be started in a soil or medium having an ideal pH range. In order to have sufficient time for the spreading, incorporation, and reaction of any needed amendments, soil samples must be collected and tested at least a few weeks before the crop is planted.

TABLE 3.1. Desirable pH ranges for several crops

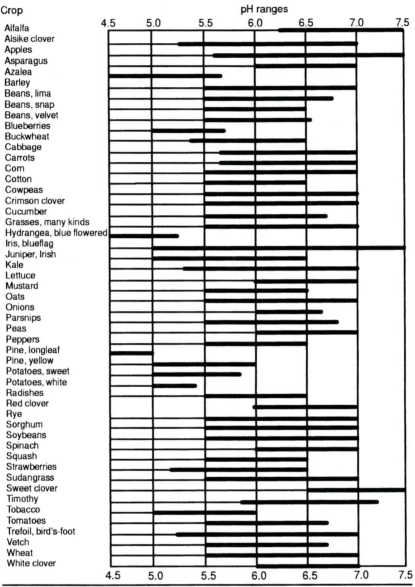

Source: Wolf, B., J. Fleming, and J. Batchelor. 1985. *Fluid Fertilizer Manual*. National Fertilizers Solutions Association, Manchester, ND 63011. Reprinted by permission of the Agricultural Retailers Association, St. Louis, MO 63146, present publishers of the *Fluid Fertilizer Manual*.

TABLE 3.2. Materials suitable for increasing soil pH and their calcium carbonate equivalents (CCE)

Liming material	Common name	Formula	CCE*
Calcium carbonate	Calcite, aragonite	$CaCO_3$	100
Calcitic limestone	High cal or calcium limestone	$CaCO_3^*$	80-100
Calcium oxide		CaO	179
Burnt lime	Quick lime	CaO^*	150-179
Calcium hydroxide		$Ca(OH)_2$	136
Hydrated lime	Slaked lime Builders lime	$Ca(OH)_2$	120-136
Dolomite		$CaCO_3 \cdot MgCO_3$	109
Dolomitic limestone	High=mg lime	$CaCO_3 \cdot MgCO_3^*$	to 108*
Calcined dolomite	Burnt dolomite	$CaO \cdot MgO^*$	to 185*
Hydrated dolomite	Hydrated dolomite	$Ca(OH)_2 \cdot Mg(OH)_2$	to 166*
Calcium silicate		$CaSiO_3$	86
Basic slag	Thomas slag	$CaSiO_3^{**}$	to 86
Blast furnace slag		$CaSiO_3^{**}$	to 86
Open hearth slag		$CaSiO_3^{**}$	to 86
Ashes, coal		variable	0-40
Ashes, wood fresh	Hardwood ash	variable***	to 80
Marl		variable	70-90
Portland cement kiln flue dust		variable****	to 100
Sugar beet lime		variable*****	80-90
Shells	Ground oyster, egg	$CaCO_3^*$	75-90

*Calcium carbonate equivalent with pure $CaCO_3$ = 100
**The slags contain various amounts of Ca, Mg, and P. Basic slag contains 27-42 percent Ca, 1-5 percent Mg, and 5-10 percent P; blast furnace slag contains 26-32 percent Ca, 2-7 percent Mg and <1 percent P; open hearth slag contains 16 percent Ca, 5 percent Mg, and 1 percent P. They also may have large amounts of micronutrients.
***Unleached wood ashes can contain 1 percent P, and 4-21 percent K; the leached ashes about 0.5 percent P, and 1 percent K_2O. Both contain various amounts of trace elements, primarily Fe.
****Can contain varying amounts of K.
*****Contains 0.05-0.6 percent P, with an average content of 0.38 percent.

TABLE 3.3. Acidifying materials and their relative capability of lowering pH as compared to sulfur

Material	S	Formula	Relative amount change as S*
	%		
Sulfur	100	S	100
Sulfuric acid (98%)	31	H_2SO_4	318
Aluminum sulfate	20	$Al_2(SO_4)_3 \bullet 18H_2O$	694
Ammonium bisulfate solution	17	$(NH_4)HSO_3$	456
Ammonium sulfate	24	$(NH_4)2SO_4$	277
Ammonium thiosulfate solution	26	$(NH_4)2S_2O_3$	304
Ammonium polysulfide solution	45	$(NH_4)Sx$	177
Ammonia sulfur solution	10	$(NH_3 + S)$	199
Iron sulfate (ferric)	17	$FeSO_4 \bullet 9H_2O$	588
Iron sulfate (ferrous)	18	$FeSO_4 \bullet 7H2O$	870
Iron pyrites (87 percent)	46	FeS_2	215
Sulfur dioxide	50	SO_2	200

* Basis of calculation: 1.8 lb. pure calcium carbonate are needed to neutralize 1 lb. NH_4-N; 3.125 lb. calcium carbonate are needed to neutralize 1 lb. S.

Source: Wolf, B., J. Fleming, and J. Batchelor. 1985. *Fluid Fertilizer Manual.* National Fertilizer Solutions Association, Manchester, ND 63011. Reprinted by permission of the Agricultural Retailers Association, St. Louis, MO 63146, present publishers of the *Fluid Fertilizer Manual.*

Timing the pH Determination

The normal change of soil pH is rather slow, so an annual pH determination from a fall sample is usually sufficient. Occasionally, use of fertilizers or water can shift the pH to an undesirable range, requiring additional amendments to correct pH during the growth of the crop. The shift is more pronounced in soils of low CEC or poorly buffered media. The need for additional pH tests beyond those collected prior to planting the crop will vary with the fertilizer program and the nature of the substrate. For hydroponics, it might be desirable to test for pH on a weekly basis, and in some cases on a daily basis. In container cultures, testing on a biweekly or even monthly basis can be sufficient. It is difficult to lay down rules as to exactly when

TABLE 3.4. Limestone rates required to change soil pH to 6.5 as affected by textural class and OM contents

Soil type	Rating	Organic matter	Soil pH				
		%	4.0-4.4	4.5-4.9	5.0-5.4	5.5-6.0	6.0-6.4
			lb/ac				
Loamy Sand	low	< 0.9	2000	1250	750	500	250
	med	0.9-1.6	3000	2000	1250	750	500
	high	> 1.6	4000	3000	2000	1000	750
Sandy loam	low	< 1.2	4000	3000	2000	1000	500
	med	1.2-2.0	5000	4000	3000	1500	1000
	high	> 2.0	6000	5000	4000	2500	1500
Loam	low	< 1.5	6000	4500	3000	2000	1500
	med	1.5-2.8	7000	5500	4000	2500	2000
	high	> 2.8	8000	6500	5000	3500	2500
Silt loam	low	< 1.8	7000	5500	4000	3000	2000
	med	1.8-3.2	8000	6500	5000	4000	3000
	high	> 3.2	9500	8000	6500	5000	4000
Clay loam	low	< 2.0	8000	6500	5000	4000	3000
	med	2.0-3.8	9000	7500	6000	5000	4000
	high	> 3.8	11000	9000	7500	6500	5500

Source: Wolf, B., J. Fleming, and J. Batchelor. 1985. *Fluid Fertilizer Manual.* National Fertilizer Solutions Association, Manchester, ND 63011. Reprinted by permission of the Agricultural Retailers Association, St. Louis, MO 63146, present publisher of the *Fluid Fertilizer Manual.*

TABLE 3.5. Approximate amounts of sulfur needed to reduce soil pH one unit

Soil texture	Sulfur
	lbs/A*
Loamy sand	300
Sandy loam	500
Loam	800
Silt or clay loam	1200

*Incorporated to plow depth

Source: Wolf, B., J. Fleming, and J. Batchelor. 1985. *Fluid Fertilizer Manual.* National Fertilizer Solutions Association, Manchester, ND 63011. Reprinted by permission of the Agricultural Retailers Association, St. Louis, MO 63146, present publishers of the *Fluid Fertilizer Manual.*

additional pH tests should be made in the field. Because of its simplicity and low cost, a pH test ought to be run whenever any other chemical tests are being made so one can evaluate the need for side dressings. An annual pH measurement is generally sufficient for perennial crops, but a soil pH test is needed, regardless of the kind of crop, whenever problems of plant growth arise.

ORGANIC MATTER

The amount and kind of soil organic matter (OM) influences the availability of several important nutrients and modifies several physical characteristics as outlined in Chapter 2.

Organic matter can contribute to the amounts of available soil N, but if large amounts of OM having a wide carbon/nitrogen (C/N) ratio are added to soils, available N can be depleted. Large amounts of surface OM can also increase gaseous losses of N from nitrogenous materials applied to the surface.

Organic matter can also release substantial amounts of other important nutrients, but is usually ties up copper (Cu). The Cu requirement of newly drained organic soils is rather high and many soils rich in OM will require some additional Cu.

Method of Analysis

Organic matter can be determined by dry combustion; Walkley-Black modification of wet combustion, or the Udy dye method, all of which are laboratory procedures. The Walkley-Black method is commonly used for rapid soil testing.

Evaluation of OM Tests

Amounts of OM associated with good growth vary according to textural class, with larger amounts being needed for the heavier soils (Table 3.6). Productivity ratings presented in Table 3.6 are probably affected to a considerable extent by the ability of OM to retain water, which may account for a less pronounced association between productivity and OM if irrigation is available.

Efforts have been made to quantify the amounts of nutrients, particularly N, released from OM, but results will vary with different soils and climates. A very rough estimate of N places the release in mineral soils in the range of 2-3 percent of the N in OM in a crop year. Estimating a mineral soil to weigh 2 million pounds per plowed acre (6 2/3 inches) and the N content of soil OM to be about

TABLE 3.6. Approximate percentages of organic matter (OM) in different soil textural classes related to productivity

Productivity rating	Sand	Sandy loam	Loam	Clay loam
		% OM		
Poor	< 0.8	< 1.2	< 1.5	< 2.0
Fair	0.8-1.5	1.2-2.0	1.5-2.5	2.0-3.0
Good	> 1.5	> 2.0	> 2.5	> 3.0

Source: Wolf, B., J. Fleming, and J. Batchelor. 1985. *Fluid Fertilizer Manual.* National Fertilizer Solutions Association, Manchester, ND 63011. Reprinted by permission of the Agricultural Retailers Association, St. Louis, MO 63146, present publishers of the *Fluid Fertilizer Manual.*

10 percent would give about 20-30 lb N released annually for each percent of OM. Such release is dependent on adequate temperature, moisture, pH, and soil aeration. In many soils, these conditions are not ideal, especially during the early growth period when lack of sufficient N can seriously limit crop production. This is the reason most growers will start the crop with considerable available N regardless of the OM level. However, it might be desirable to withhold some N if the soil OM is in a good to high range, making additions only if the OM fails to release sufficient N by a critical time.

Frequency of Testing

The percentage of OM changes rather slowly, and an annual or preplanting determination is sufficient.

CATION EXCHANGE CAPACITY

The capacity of the soil to hold cations that can be readily exchanged for other cations is known as the *cation exchange capacity (CEC)*. It is expressed in terms of milliequivalents (meq) per 100 grams (g) of dry soil.

Cation exchange involves the exchange of one positively charged ion for another. The positively charged cations of the soil solution are attracted and held by the essentially negatively charged *colloidal* (very fine) particles consisting of clay and humus. (The negatively charged anions are also attracted, but to a much smaller degree.) The cations hydrogen (H+), ammonium (NH^{4+}), potassium (K^+), calcium (Ca^{2+}), magnesium (Mg^{2+}), sodium (Na^+), and aluminum (Al^{3+}) make up the major portion of cations held by the colloidal material, but small amounts of copper (Cu^{2+}), iron (Fe^{2+}, Fe^{3+}), manganese (Mn^{2+}), and zinc (Zn^{2+}) are also held.

Textural Class and CEC

The CEC varies with textural class due to the dependence of CEC upon the amounts of colloidal materials (clay and humus). Soils with more clay and/or humus will have higher CEC as reflected in Table 3.7.

Effects of Clay Type

The CEC is affected not only by the amounts of clay and OM, but also by the type of clay. Hydrous oxides have a CEC of 2-6 meq/100 g; kaolinite clay 3-15 meq/100 g; illite 10-40 meq/100 g; montmorillonite 80-150 meq/100 g; and vermiculite 100-150 meq/100 g of soil. Highly weathered soils with predominantly kaolinitic-type clays can be expected to have a much lower CEC than soils with about the same clay content but with montmorillonite- or vermiculite-type clays.

Base Saturation

Percent base saturation denotes the percentage of the total CEC that is occupied by base cations (i.e., cations capable of reducing acidity) such as potassium (K^+), sodium (Na^+), magnesium (Mg^{2+}), and calcium (Ca^{2+}). Thus, soils with low base saturation are also very acid soils. The percent base saturation affects pH, nutrient availability, and leaching.

Plants tend to remove base cations, replacing them with H+, an acid cation, increasing the acidity of the soil. Availability of the cations tends to increase as their percent base saturation increases,

TABLE 3.7. Typical cation exchange capacities (CEC) of several soil textural classes

Textural class	CEC
	meq/100g
Sand	<4
Sandy loam	3-10
Loam	8-18
Silt loam	10-25
Clay and clay loam	20-50
Organic soils	50-200

provided that certain ratios of the cations are not exceeded. Ideal percentage base saturations of the major cations are about 65 to 75 percent for Ca, 10 to 15 percent for Mg, and 3 to 7 percent for K. Attempts to achieve these ideal percentages by using fertilizer or lime have not necessarily provided for better crops. Apparently, soils that have these ideal percentages produce good crops, but good crops can be grown without ideal percentages provided that sufficient nutrients are added to favorably affect the soil solution adjacent to the roots.

Importance of CEC and Base Saturation Measurements

Cation exchange data provide an index of the amounts of cations that are held strongly enough to slow losses from leaching or volatilization, but are still readily available for plant uptake. The CEC also affects the efficiency of several herbicides, with larger amounts required on soils with elevated CEC in order to obtain satisfactory weed kill.

Much larger amounts of cations (NH^{4+}, K^+, and Mg^{2+}) can be applied to soils with elevated CEC with little danger of leaching or volatilization losses. Unfortunately, NH4+ can be converted to the more leachable anion, nitrate (NO_3^-), and be lost unless nitrifying inhibitors are added. Conversely, small amounts of cations need to be added to soils with low CEC in order to minimize leaching losses, making it mandatory to use one or more sidedressings on light soils if leaching is a problem.

The beneficial effect of adequate CEC in retaining cations is offset to some extent by the need to add much larger amounts to obtain sufficient availability if levels are low. The percent base saturation can be helpful in estimating lime and fertilizer requirements. The presence of high Na saturation (greater than 10 percent) may limit the type of crop that can be grown or require special treatments to alleviate the problem.

Methods of Analysis

An estimate of CEC is possible from CEC data of textural classification (see Table 1.3 in Chapter 1), but more accurate determinations are usually worthwhile. Several laboratory procedures provide much more reliable data. A common procedure employs ammonium acetate solution to leach the soil so that all exchangeable cation sites are occupied by ammonium. The excess ammonium is removed by leaching with alcohol. The ammonium held by the soil is displaced by another cation and is determined as ammonia, usually by titrating with standard sulfuric acid. The CEC is equal to the milliequivalents of displaced ammonium. The individual cations used to calculate percentage base saturation are determined separately in the original leachate. Annual or preplanting tests for total CEC and base saturation are usually sufficient.

CONDUCTIVITY AND SOLUBLE SALTS

Conductivity of a soil solution is proportional to the amounts of soluble salts. Conductivity measurements are helpful in selecting suitable crops and fertilizer, amounts and timing of fertilizers and water applications, and the manner in which planting beds are to be constructed.

Conductivity measurements of some soils are also useful in indicating whether there is sufficient N and K for good crop growth. In noncalcareous, low-sodium soils, conductivity readings of an extract made from two parts of water to one part soil can be used for a quick diagnosis of N and K. Readings of 0.5-1.0 millimhos (mmhos) are usually indicative of sufficient amounts for rapid growth. Readings of 0.2-0.5 mmhos show sufficient conductivity for growth maintenance, and conductivities of less than 0.2 mmhos indicate the need for supplements. If previous soil tests have shown the presence of ample K, then only additions of N are necessary. In most sandy soils or media low in K, both N and K need to be added. Such indications are preliminary in nature and need to be confirmed occasionally by tests for the elements, but conductivity tests can be used in the field to get a quick reading on the availability of fertilizer for the crop.

The ability of conductivity tests to help diagnose the presence of excess salts and inadequate levels of N and/or K make them invaluable to a grower. The test is simple, rapid, and can be run easily in the field or in a small farm laboratory. The cost of meters capable of running the test start at about $50, with more advanced models in the $500 range. No grower or consultant should be without a meter, and no crop should be started or fertilizer applied before taking a conductivity reading of the soil. Conductivity readings need to be taken at least once a year. Ideally, a test should be conducted before each application of fertilizer, and certainly if conditions warrant troubleshooting.

Sampling Soils

Normal soil samples collected as outlined in Chapter 1 are suitable for most conductivity measurements. Crops in the early stage of growth that are grown in soil covered by plastic require special samples. As water evaporates from openings in the plastic, soluble salts are left at or near the soil surface, at times rising to concentrations high enough to cause serious problems for the developing seedling. Conductivity measurements of a sample collected in the upper inch of soil close to the developing plant will reveal whether salts are a problem. If salt levels are high, concentrations can be lowered to safe levels by application of water directed to the openings in the plastic. Salt accumulation at the soil surface in the plastic opening seldom is a problem once the plant is large enough to shade it because the shade reduces water evaporation from the opening. Less salts are brought to the surface as the reduced amount of water evaporates.

Conductivity Measurements

Conductivity or specific conductance can be measured in a 1:2 soil to water extract or in a saturated paste. The conductivity of the saturated paste extract is a more reliable gauge, but unlike the soil-water extract, its determination is confined to the laboratory.

Soil-Water Ratios

Various ratios of water to soil have been used to obtain the water extract, the most common being 5:1 or 2:1 water to soil, with recent tendencies toward the narrower ratio. To determine conductivity of a water extract, two parts of water are added to one part of well-mixed soil sample (by volume). Deionized or distilled water is preferred, but tap water can be used as long as its conductivity is later subtracted from that determined by a soil test.

The soil and water are mixed well with a clean spatula or spoon, and allowed to stand with intermittent mixing for one hour. The temperature setting of the conductivity meter is adjusted to the temperature of the supernatant liquid in the mix and conductivity is determined by placing the conductivity probe in the supernatant liquid or in a filtered extract of the sample.

In-Soil Measurement

Measurement of salts can be made directly in the soil by porous matrix salinity sensors or four-electrode units. This type of measurement is useful for irrigation monitoring and determining drainage needs. The porous matrix salinity sensor measures salinity of soil water that has been imbibed into a 1 mm thick ceramic disc. The four-electrode unit measures the resistance to the flow of an electric current between two electrodes placed in the soil while current between a second pair of electrodes is passed through the soil. A constant current is passed through the outer pair of electrodes via electrolytic conduction while a resistance meter is used to measure resistance to current flow between the inner pair of electrodes. For any given soil type, electrical conductivity can be correlated to resistance between electrodes if soil water is constant. Conductivity is calculated by the following equation:

$$Ec = 1/(2 \, \pi aR) \cdot 1000$$
Where Ec = mmhos/cm
π = 3.1417
a = distance between electrodes in centimeters
R = resistance in ohms

In practice, conductivity can be measured in the root zone of an area by placing the electrodes equidistant in a straight line and inserting them about one inch in the soil before taking measurements. Conductivity is measured in different areas by moving the probes. On spot measurements can be made with a single probe device that has the four electrodes inlaid in a single shaft. The single probe device measures conductivity in a smaller volume of soil than the separate four electrode unit but allows for faster evaluation of large areas due to easier handling.

Expression of Results

Electrical conductivity (EC) is usually expressed as millimhos per centimeter (mmhos/cm) although there has been a tendency to use decisiemens per meter (ds/m). Some relationships of the various expressions for conductivity are as follows:

MHO = reciprocal ohm
mhos/cm = 1000 millimhos/cm (mmhos/cm)
mmhos cm = 1000 micromhos/cm (μmhos)
siemen per meter (S/m) = 10 mmhos/cm
decisiemens per meter (dS/m) = mmhos/cm
mmhos cm = 0.1 S/m

Other useful relationships with conductivity are:

Osmotic pressure (OP) = $-0.36 \times$ EC in mmhos
Parts per million (ppm) ions = $640 \times$ EC in mmhos
Milligrams per liter (mg/1) of ions = $640 \times$ EC in mmhos
Milliequivalents per liter (meq/1) of ions = $10 \times$ EC in mmhos

Interpretation of Results

The interpretation of the results is dependent upon the method of preparing the extract. Much higher values can be tolerated if a saturated extract or lower water to soil ratios are used. Typical interpretations for two parts water to one part soil or five parts water to one part soil or saturated paste extracts are given in Table 3.8. Higher values than those given in this table are acceptable with plants tolerant to salts or if the soil contains large quantities of OM and moisture.

Frequency of Determination

Conductivity readings of field soil need to be taken at least once a year. Ideally, a test should be conducted before each application of

TABLE 3.8. Interpretation* of soluble salt readings

Saline class	Rating	Water extract by volume		Saturated paste extract	Effect on crop
		1 soil: 2 water	1 soil: 5 water		
			mmhos/cm		
none	too low	< 0.15	0.08-0.15		lacks nutrients**
none	low	0.15-0.50	0.15-0.25	< 2	negligible***
weak	OK	0.51-1.50	0.30-0.80	2-4	slight****
moderate	high	1.51-2.25	0.80-1.00	4-8	many crops restricted*****
strong	excess	> 2.25	1.0-1.5	> 8	use only tolerant crops

*Interpretation is based on normal moisture content. High organic matter in soil increases tolerance by increasing amount of moisture held.
**Acceptable for seed germination but too low for seedlings. The amount of available N and K is probably too low for growth of most plants.
***Salts are satisfactory for most plants but lower end of scale may indicate too little N and K for rapid growth.
****The upper levels are too high for sensitive plants and for most plants during germination.
*****Sensitive and even moderately sensitive plants can be severely restricted.

Source: Based on the author's observations and data presented by Waters, W. E., J. NeSmith, C. M. Geraldson, and S. S. Woltz. 1972. The interpretation of soluble salt tests and soil analysis by different procedures. *Florida Flower Grower* 9(4).

fertilizer, and certainly if conditions warrant troubleshooting. Measurements of artificial soils or hydroponic media need to be made much more frequently. Although hard-and-fast rules are difficult to formulate, monthly evaluation of artificial soils probably is adequate, whereas weekly determinations of hydroponic media would be more suitable.

NUTRIENTS

A minimum of 16 chemical elements are essential for the growth and reproduction of green plants. These elements and their chemical symbols are: carbon (C), oxygen (O), hydrogen (H), nitrogen (N), phosphorus (P), potassium (K), calcium (Ca), magnesium (Mg), sulfur (S), boron (B), chlorine (Cl), copper (Cu), iron (Fe), manganese (Mn), molybdenum (Mo), and zinc (Zn). Air and water supply needed C, H, and O, and the air may be a source of some N and S. The other elements, and most of the N and S, are primarily derived from the soil. Insuring adequacy of these elements by measuring their availability in the soil offers a very practical means of controlling crop growth.

In addition to these essential elements, there are at least three elements that can cause serious problems of plant growth if present at elevated levels. Aluminum, although tolerated by and even beneficial to some plants, can in relatively small quantities seriously reduce production of many crops. Because its availability is reduced by elevated pH, problems seldom exist in soils with pH values of greater than 6.5. Excess Na can adversely affect soil structure and increase salts to damaging levels. The Na problem is restricted primarily to alkaline soils. Fluoride (F) in small quantities can affect the production of several plants. Ornamentals are particularly prone to losses, since leaf lesions resulting from excess F seriously reduce their value. Problems are primarily present in acid soils.

Testing for Available N

Despite the importance of N in crop production, most soil laboratories have shied away from determining available N [ammonium-nitrogen (NH_4-N) + nitrate-nitrogen (NO_3-N)]. The reason usually given is that the amounts tend to change too quickly. Rather than make recommendations based on such transitory information, some laboratories have used soil OM contents or nitrogen released by incubating soil samples in the laboratory as a basis for measuring potential soil N release. Both methods, although helpful, lack accuracy as to amounts or timing of N released by the soil under actual conditions. As a result, crop potentials often have not been obtained or there has been some nitrate pollution of ground waters due to excessive use of N or incorrect timing of the N application.

Testing for available N using samples collected in the fall in humid zones is largely meaningless, as N levels at planting time will be very low and can be largely disregarded. This is not true of soils in arid or semi-arid zones, where enough N can accumulate in top and subsoils to appreciably influence the succeeding crop. Nor is it true for samples collected during the late spring or summer from soils in humid zones that will be used for late spring plantings or a second crop. In all of these cases, tests for available N may reveal the presence of enough N to allow for a substantial reduction in the N application.

Even in the case of early spring plantings on humid-zone soils, available soil N tests can be useful, but the tests need to be run after the crop is established. The OM content or N released by incubation from soil samples collected prior to planting can be used as a preliminary guide for N recommendations, but only a portion of the total N suggested ought to be applied at planting. The remainder of the N or any additional amounts can be applied if and when later periodic tests for available N show a need. Adjusting applications to match need can greatly increase N efficiency while reducing the chances for pollution of ground waters.

In recent years, there has been greater recognition of the need for available N tests. An early spring N test of soil to a depth of two feet has been used in a few states to estimate the N needs of corn and wheat grown in less humid areas. In more humid areas, a pre-side-dress test of samples collected to a depth of one foot is being used to determine the need for sidedressing N. No N is applied if soil N is greater than 20-30 ppm when corn is at a height of 12-18 inches. If soil N is less than 20 ppm, the need for sidedress N is often confirmed by a test for NO_3 in stalk tissue.

Evaluation of all the elements important for crop production must be made. Evaluations can greatly help in formulating suitable fertilizer recommendations and provide invaluable information for troubleshooting.

Sample Frequency

The frequency of running these tests is dependent on such factors as: (1) mobility of the element; (2) type of soil; (3) extent of leaching; and (4) kind of crop. The old recommendations of sampling every one to three years were made by practitioners who failed to make full use of soil testing. Once-a-year sampling may be satisfactory for the relatively immobile elements (P, K, Ca, S, Cu, Mn, and Zn). It may be all that is needed also for the mobile elements (N, K, and Mg) under arid or semi-arid conditions, or with woody perennials grown in the field. But once-a-year sampling is not sufficient to evaluate the need for mobile elements under humid conditions. Nor is it satisfactory for plants grown on light or artificial soils in restricted containers subject to leaching. Several extra samples may be necessary to evaluate the need for mobile elements in the field, and samples collected every two to four weeks may be needed to fully evaluate the nutrition of crops grown in containers subject to leaching.

More frequent testing for pH, conductivity, and the nutrient elements may be required to maintain desirable growing conditions for hydroponically grown plants. M. Schwarz of the Negev Institute for Arid Zone Research in Israel recommends testing every two days for pH; 15-30 days for osmotic pressure; 10-15 days for NO_3-N, 10-20 days for NH_4-N and PO_4-P, 20-30 days for K; every 20 days for Fe; and every 30 days for Ca. While Fe and Ca levels should be tested routinely, the levels of these elements, as well as Mg, SO_4-S, Na, and Cl, need to be determined whenever shortages or excesses (as indicated by plant symptoms) are suspected. More frequent or routine testing for osmotic pressure, PO_4-P, Ca, Mg, SO_4-S, Fe, Cl, and Na is suggested if saline waters are used.

Timing the Sample

The usual timing for sampling in the Northern Hemisphere is late summer or early fall. This is a very satisfactory period because soil conditions for sampling are usually excellent; there is sufficient time before the next planting for running the samples and ordering soil amendments; and plant samples can be collected at the same time allowing for a complete evaluation of fertility practices.

This timing period designed for the next crop may not adequately provide the most desirable fertilizer program, because the tests may fail to show whether the previous applications were sufficient for the previous crop. If nutrients are found to be low, it is not always possible to pinpoint when they first fell below the desirable levels. For such information, many more samples during the crop season or crop logging are needed. Another criticism of fall sampling is that conditions may change by the time planting is made. This criticism is perhaps valid only for N, and only in regions where leaching is limited. In such regions, sufficient N can carry over for the next season and should be evaluated shortly before the crop is planted.

Available N in the form of NH_4-N and NO_3-N needs to be determined for replant crops, for spring-planted crops in dry regions, and for sidedressings of N.

Although problems of inadequate nutrients are usually associated with N, additional tests for K and Mg during the growing season are often needed to fully evaluate fertility programs. The need to obtain additional N, K, and Mg data while the crop is growing is greatest for annual crops grown on light soils in humid regions. There is much less need for tests on the heavier soils, particularly if ammoniacal fertilizers with nitrifying inhibitors are used.

There appears to be little value in frequent sampling of soils on which woody perennial crops are grown, except in the first year or two as they are being established. Many of these established crops have reserves of N and other elements in their tissues, making the plant less affected by temporary shortages in the soil. A single sample in the late summer or early fall is usually sufficient for most established fruit crops.

Even when evaluating the adequacy of fertilizer recommendations, with later samplings, these generally need to be completed by the time the crop is half grown. Seldom are yields affected by later fertilizer applications if nutrients are adjusted to correct any shortages noted in the sample collected at mid-growth. In a few cases, late applications of N and K can affect the appearance of some high-priced crops enough to warrant sampling up to shortly before harvesting. Late sampling just prior to shipping potted plants can be helpful in adjusting nutrients to levels desirable for particular holding and lighting conditions.

Sample Handling

It is important to minimize the release of available N from OM as cold soil samples are placed in warm rooms. Filling and compacting the soil in the container while closing it tightly will limit nitrification, so that no other procedures will be necessary if samples are analyzed within a few days. Samples that are to be kept for some time before analysis should be stored in refrigerated areas.

Field versus Laboratory Analysis

Obtaining nutrient analysis directly on the farm can save a few days of time. Several kits suitable for farm operations are available for macronutrient evaluation and some even for micronutrients. There is some question as to their accuracy because solutions deteriorate in time and conditions under which the tests are developed are often poorly controlled. Additional problems result from translating results into fertilizer recommendations because of limited experience.

Such objections to on-farm analyses are eliminated if a true laboratory is established, the tests are run under controlled conditions, and the person making the recommendation has sufficient experience. Generally, such laboratories are restricted to fairly large operations.

Most growers will have to rely upon outside laboratories for the major portion of their analysis. Kits capable of running at least N and K tests can be helpful in an emergency, but kits have to be maintained by renewing test solutions periodically. Whatever information is obtained by a test kit needs to be corroborated by submitting duplicate samples to a qualified laboratory. Several sources for test kits are presented in Appendix 5.

Both private and governmental laboratories can provide rapid soil analysis and make the necessary recommendations for soil amendments. Governmental laboratories,* usually associated with State Extension Services, can provide the services at lower costs but results usually are not as rapid.

Extracting Solutions

Analysis of the soil solution can provide answers as to the sufficiency of nutrients, but there is no simple method of extracting the soil solution. Rather, various solutions are employed to extract nutrients, the analyses of which need to be correlated with plant growth. Some of the more common extracting solutions being used for this purpose are given below.

Morgan

The solution introduced by M. F. Morgan of Connecticut in the 1930s uses 0.75 N sodium acetate buffered at pH 4.8 with acetic acid. It was one of the first "universal" solutions, permitting determination of most significant nutrients in a single extraction. It is well adapted to the acid, low CEC soils. A modification introduced by the author which adds DTPA (0.18 g/l) gives better correlations with the micronutrients. The modified Morgan (Morgan-Wolf) is well adapted for nutrient evaluation of field or greenhouse soils that are noncalcareous and of low to moderate CEC (less than 18 mg/100 g).

Mehlich I

This double acid solution consisting of 0.05 N hydrochloric acid and 0.025 N sulfuric acid is well adapted to acid, sandy soils with low CEC, but has been largely superceded by Mehlich III.

---

*A detailed list of soil testing laboratories and the services they provide is available from Micro-Macro Publishing Co., 183 Paradise Blvd., Athens, GA 30607.

Mehlich III

The solution consists of 0.2 N acetic acid, 0.25 N ammonium acetate, 0.015 N ammonium fluoride, 0.013 N nitric acid, and 0.01 M EDTA, and is now being used extensively in the Eastern seaboard states for multielement extraction of neutral and acid soils. Because of its considerable buffering capacity, it has also proven useful for extracting basic soils in these and other states.

Ammonium Acetate and Bray-Kurtz or Olsen Extracting Solution for P

Several solutions are used for evaluating nutrients from soils of medium to high CEC. The methods have been used extensively in the Great Plains states. At least one commercial laboratory is using them for a wide range of soils in the continental U.S. and one location in Canada. The CEC and the cations K, Ca, Mg, H, and Na are determined by extracting the soil with 1 N ammonium acetate solution buffered at pH 7.0. Bray-Kurtz #1 solution consisting of 0.03 N ammonium fluoride and 0.025 N hydrochloric acid is used for determining P in acid soils. Olsen extracting solution consisting of 0.5 percent sodium bicarbonate solution (see number 5 below) is used to determine P in alkaline soils. Micronutrients are extracted by using 0.1 N hydrochloric acid for noncalcareous soils, a solution of 0.005 M DTPA + 0.1M TEA and 0.01 M calcium chloride for a wide range of soils, or a solution of 1 M ammonium bicarbonate + 0.005 M DTPA adjusted to a pH 7.6 for alkaline soils. Nitrogen is extracted by another solution, such as sodium chloride, potassium chloride, or calcium sulfate.

Olsen

This solution consisting of 0.5 M sodium bicarbonate adjusted to a pH of 8.5 was first used to determine P in alkaline and calcareous soils. It has since been modified by P. N. Saltanpour of Colorado by including 1.97 g DTPA per liter (AB-DTPA) for the additional and simultaneous extraction of K, Cu, Fe, Mn, and Zn.

Spurway

A solution of 0.018 N acetic acid is used extensively for extracting several nutrients from potted soils used for growing flowers and bedding plants. This method of extraction can be useful if repeated frequently and fertilization consists mainly of water soluble ingredients. Drawbacks of the acid alone are that it fails to measure the exchangeable ions which can contribute to the overall fertility. The failure to measure the exchangeable ions is not a serious problem if the procedure is repeated frequently enough, i.e., once a week. In working with greenhouse crops, we have found that our modified Morgan method of extraction based on soil volume extraction was more useful than the Spurway method.

Saturated Media

Water is used to saturate the soil and the extract is collected by vacuum. This method, developed at U.S. Salinity Laboratory, has been used effectively for greenhouse soils in Michigan and by Dr. Geraldson for the poorly buffered soils of Florida. By determining the relationship of the extracted nutrient to the total soluble salt content, Geraldson has developed an "intensity and balance" system to evaluate the suitability of a particular concentration. In this system, suitable percentages of total soluble salts for several nutrients are: 3-10 percent NO_3-N, 11-13 percent K, 14-16 percent Ca, and 4-6 percent Mg. Sodium and Cl each should be less than 10 percent of the total soluble salts.

Water

Water suspensions shaken for different periods of time or water poured through container media has been used to extract easily soluble nutrients.

Water extraction of container soils has the advantage of not rupturing slow-release fertilizer granules commonly used in these soils. Water extraction, however, fails to determine the reserves of several nutrients and, therefore, is usually not suitable for well-buffered soils, unless the tests are repeated rather frequently. The water-soluble P test is helpful in determining the short-term need for extra P in these soils at times. Such soils may respond to extra P at planting if soils are cold or the applied P is too far from developing roots, despite the fact that they might have ample P as measured by Bray-Kurtz #1 and #2 or modified Morgan.

Egner

The solution of 0.1 N ammonium lactate and 0.4 N acetic acid, which has had limited use in the U.S., has been more readily exploited in Europe.

EUF

The method combines both electrodialysis and ultrification to simultaneously extract NH_4-N, NO_3-N, P, K, Ca, Mg, Na, S, B, Mn, and Zn. A soil suspension (one part soil: ten parts water) is added to a special apparatus consisting of three compartments, separated by microfiber filters and platinum electrodes while voltage is applied. A vacuum applied to the outer cells allows for the withdrawal of dissolved elements which are determined in the removed solution. This method of extraction is primarily used in Europe.

Choosing a Suitable Method of Extraction

The selection of an extraction method needs to be based on several soil characteristics. The modified Olsen is best suited for high-pH soils, whereas several of the other methods can be used for acid soils. Mehlich III seems to work well for soils with a wide range of pH values. If testing only for micronutrients, the ammonium bicarbonate DTPA can be used for the high-pH soils and the 0.1 N HCl for acid soils. The DTPA, TEA, calcium chloride solution can be used on soils of varying pH. The Bray-Kurtz #1, Egner, modified Morgan, Mehlich I, and Spurway (water extraction) are better designed for soils with low to medium CEC. Ammonium acetate (pH 7.0) and Bray-Kurtz #2 are more suitable for soils with medium to high CEC. Although EUF is suitable for a wide range of soils, the cost of units and the length of time for extraction make this system more costly than some other methods.

Comparison of Test Results

Often, it is not possible to quantitatively compare results of one laboratory with that of another because they may be using different extraction methods or using different methods of storing soil samples, preparing them, or shaking them for extraction. Even recommendations for fertilizers may be different after analyzing duplicate samples because of different yield goals or philosophies regarding fertilizer use. For these reasons, it is preferable for the grower to seek out a reliable laboratory and remain with the same laboratory for a number of years. Close cooperation between grower and laboratory director can usually refine the results so that yields are increased over time.

Sample Weight versus Volume

There has been some difference of opinion as to whether the soil sample ought to be weighed or measured by volume. Volume measurement appears to be more reliable as it corresponds to a volume of soil being explored by plant roots. The use of volume is particularly useful if many different types of soil are tested or there is appreciable difference in soil specific gravity. For many soils, a factor of 0.227 converts ppm wt/vol into lbs/a per each inch of sampling depth. The factor 1.0 needs to be used for a 4.4-inch depth; 1.5 for 6 2/3-inch depth (common plow depth); and 1.82 for an 8-inch sampling depth.

Methods of Analysis

Several different types of analytical procedures can be used. Colorimetric and turbidimetric methods were used almost exclusively in the early days of rapid soil analysis. Some of these procedures are still in use today, particularly for N, P, and B. Flame emission became the preferred method for K, Ca, Mg, and Na, and atomic absorption for Cu, Fe, Mn, and Zn, although atomic absorption has been effectively used for all of these elements. Ion electrodes are being used to some extent for analysis of NO_3-N, NH_4-N, K, Ca, and F. Recently, plasma instruments, capable of measuring a number of significant elements simultaneously and rapidly, have been used more extensively, although instrument cost has limited the use of plasma.

Usually only NO_3-N is measured in evaluating the need for N, but in many situations both NH_4-N and NO_3-N need to be determined. Although NH_4-N is usually converted rapidly in well-aerated, moist, warm soils to NO_3-N, full conversion may not take place if soil conditions are unfavorable for microbial action. The presence of nitrifying inhibitors, added at times to prevent N losses, can greatly slow down the process. On soils that have received large quantities of certain insecticides, herbicides, and fumigants over the years, microbial action also may be limited, allowing for long-term persistence of any added NH_4-N.

Tests for NH_4-N and the NO_3-N can be run separately on sample extracts and the total N used as an index of available N. Alternatively as outlined in the Morgan-Wolf procedure, the NO_3-N in the soil extract can be first reduced to NH_4-N by use of titanous chloride and the NH_4-N determined by Grave's or Nessler's reagent. The NH_4-N can be determined by ion electrode, indophenol blue reaction, steam distillation with magnesium oxide, or by microdiffusion with magnesium oxide. Nitrates can be determined by nitrate electrode, colorimetric procedures, or ion chromatography.

Expression of Results

Results are usually expressed as ppm or pounds per acre. Normal conversion of ppm to pounds per acre for mineral soils can be made by multiplying ppm by a factor of two.

Ratings of Soil Test Results

Results also can be categorized as Very Low (VL), Low (L), Medium (M), High (H), and Very High (VH) based on average performance. These categories are not absolutes, but rather are intended to give the grower some idea as to the adequacy of the amounts determined for sustaining growth and the need for fertilizer. While different systems may use somewhat different terms and evaluations, the categories have meanings somewhat related to those given in Table 3.9.

TABLE 3.9. Ratings for soil nutrient tests

Rating	Yield potential and need for fertilizer
Very Low or Deficient	Very poor—less than 50% of crop's potential expected. May show element deficiency. Needs full fertilizer amount.
Low or Poor	Low—50-75% of crop's potential. No sign of deficiency. Needs generous fertilizer application.
Medium or Sufficient	75-100% of crop's potential without fertilizer addition. Fertilizer needed for maintenance and/or quality for high value crops.
High	Very good with little or no yield increase to added fertilizer expected.
Very High or Excessive	Crop response may be affected adversely. No favorable response to fertilizer.

General Ratings

General ratings as given in Table 3.9 are based on correlations of nutrients extracted with yields of many crops. The nutrient levels that give these responses will vary with the method of extraction, kind of crop, nature of the soil, and sufficiency of water (irrigation). Generalized ratings for several methods are given in Table 3.10.

Ratings Based on Crop Requirements

Crops vary greatly as to their nutrient requirements. The differences in need can be adjusted by altering the rating of the soil test or varying the rate of fertilizer to correspond to the plant's needs. Several states have variable ratings for Bray-Kurtz P and Mehlich tests for P, K, Ca, and Mg. Many practitioners use a general rating but vary the fertilizer application to correspond to crop needs at various levels.

TABLE 3.10. Levels for categorizing ratings of several nutrients obtained with different extracting solutions

Extractant		VL	L	M	H	VH
N						
AB-DTPA	pp2m		0-20	21-40	41-60	> 60
Morgan (Wolf)						
(NO3+NH4)	pp2m	0-24	25-49	50-149	150-400	> 400
Sat. ext. (NO3)	pp2m	0-79	80-159	160-279	280-400	> 400
Spurway (NO3)	pp2m	0-9	10-40	41-99	100-120	> 120
(NH4)			< 2	2	5	10
Water soluble						
(NO3)	pp2m		10	20	50	
(NH4)	pp2m			2	10	25
P						
AB-DTPA	pp2m		0-6	7-14	15-22	> 22
Bray-Kurtz #1	pp2m	0-11	12-26	27-39	40-58	58
Bray-Kurtz #2	pp2m	0-14	15-39	40-79	80-118	> 118
Mehlich #1*	pp2m	0-13	14-26	27-45	46-90	> 90
Mehlich #2*	pp2m		< 38	38-53	54-90	> 90
Mehlich #3*	pp2m	0-20	21-60	61-120	121-240	> 240
Morgan (Wolf)	pp2m	0-9	10-24	25-100	101-160	> 160
Olsen	pp2m		0-11	12-24	> 24	
Sat. extract	pp2m	0-5	6-11	12-20	21-38	> 38
Spurway	pp2m	0-5	5-9	10-19	20-30	> 30
Water soluble	pp2m	0-1	1.1-2.9	3.0-20	> 20	
K						
AB-DTPA	pp2m		0-120	121-240	241-360	> 360
Ammon. Acet.	% sat.	0-2.9	3-6.9	7-9	>9	
Mehlich #1*	pp2m	0-37	38-73	74-123	124-248	> 248
Mehlich #3	pp2m	0-32	33-96	97-195	196-384	> 384
Morgan (Wolf)	pp2m	0-49	50-99	100-350	351-500	> 500
Sat. extract	pp2m	0-99	100-219	220-359	360-520	> 520
Spurway	pp2m		< 20	20-59	60-100	> 100
Water soluble	pp2m		10	20	40	
Ca						
Ammon. Acet.	% sat.	< 50	50-69	70-80	> 80	
Mehlich #1*	pp2m*		0-175**	176-300	> 300	
Mehlich #3*	pp2m	0-469	470-705	706-1100	> 1100	
Morgan (Wolf)	pp2m	0-499	500-999	1000-5000	> 5000	
Sat. extract	pp2m	0-139	140-279	280-439	440-650	> 650
Spurway	pp2m		80	200	300	400

		VL	L	M	H	VH
				Mg		
Extractant		VL	L	M	H	VH
Ammon. Acet.% sat.		< 5	5-9	10-15	> 15	
Mehlich #1	pp2m		0-29**	30-60	> 60	
Mehlich #3	pp2m	0-55	55-90	91-155	> 155	
Morgan (Wolf)	ppm2	0-49	50-99	100-1000	> 1000	
Saturated extract	pp2m	0-59	60-119	120-199	200-300	> 300
Spurway	pp2m	0-2		10		
				B		
AB-DTPA	pp2m				> 10.0	
Hot water	pp2m	0-0.7	0.8-1.5	1.6-2.5	2.6-4.0	> 4.0
Morgan (Wolf)	pp2m	0-0.4	0.5-0.9	1.0-2.0	2.1-4.0	> 4.0
				Cu		
AB-DTPA	pp2m		0-0.4	0.5-1.0	>1.0	
DTPA	pp2m	0-0.5	0.6-1.7	1.8-2.5	2.6-5.0	> 5
0.1 HCl	pp2m	0-0.5	0.6-1.7	1.8-3.0	3.1-6.0	> 6.0
Mehlich #3	pp2m	0-0.2	0.3-1.0			
Morgan (Wolf)	pp2m	0-0.9	1.0-2.4	2.5-10.0	10.1-20	> 20.0
				Fe		
AB-DTPA	pp2m		0-6.0	6.1-10.0	>10.0	
DTPA	pp2m	0-4	5-20	21-33	34-50	> 50
0.1 HCl	pp2m	0-7	8-23	24-49	50-100	> 100
Morgan (Wolf)	pp2m	0-2.5	2.6-4.9	5.0-50	51-100	> 100
				Mn		
AB-DTPA	pp2m		0-1.0	1.1-2.0	> 2.0	
DTPA	pp2m	0-7	8-17	18-25	26-60	> 60
0.1 HCl	pp2m	0-11	12-29	30-59	60-100	> 100
Mehlich #3	pp2m	0-2.6	2.7-8.0			
Morgan (Wolf)	pp2m	0-2.5	2.6-5.0	5.1-20.0	20.1-40.0	> 40
				Zn		
AB-DTPA	pp2m		0-1.9	2.0-3.0	>3.0	
DTPA	pp2m	0-0.9	1.0-1.9	2.0-5.9	6.0-12	> 12
0.1 HCl	pp2m	0-2.0	2.1-5.9	6.0-10.0	10.1-16	> 16
Mehlich #3	pp2m	0-0.6	0.7-2.0			
Morgan (Wolf)	pp2m	0-2.4	2.5-4.9	5.0-50	51-100	> 100

*Amounts for the ratings can vary with different soils and crops.
**These are some of the lowest values for these ratings.

Ratings Based on Soils

The rating of a particular test may also be dependent on the type of soil. The ratings of the cations are particularly affected. Cation availability is influenced by the percent saturation, which is affected by total CEC; the CEC is related to clay content–a dominant characteristic determining soil type. The amounts of N released in a crop year may also be associated with soil type, making the ratings of N dependent to some extent on the soil type, levels of OM, and available moisture. The variable effect of soil on the availability of several nutrients can be taken into account by using a common rating but varying the fertilizer application for the different soil characteristics. The ratings can also be varied corresponding to soil conditions. In Kansas, ratings for N availability are modified according to moisture status. The ratings for P, K, Ca, Mg, and B in the Morgan (Wolf) method are varied with different soil textural classes (see Table 3.11).

Fertilizer Recommendations Based on Soil Analysis

A general idea of crop nutrient needs can be gained from the amounts removed by good yields of several crops as given in Table 3.12. In addition to nutrients removed by crops, some nutrients are lost as a result of erosion, leaching, fixation, and volatilization. But the soil is capable of supplying nutrients. Both the losses and amounts supplied by the soil vary considerably with crop, climate, soil, type of fertilizer, and the manner in which the fertilizer has been applied.

Soil tests, by defining the character of the soil and indicating the amounts of nutrients available for the crop, can suggest the most appropriate amounts of fertilizer, types of fertilizer, and the means of applying them. But the actual amount of fertilizer additions indicated by any particular soil test value will vary with such factors as the crop, yield goals, climate, type of soil, soil moisture, soil pH, depth of soil, previous crop, use of manures, cover crops, and other organic materials.

In evaluating ratings for soil tests, test results are usually correlated with crop yields in a given region on certain soils of similar

TABLE 3.11. Ratings of nutrient levels (ppm) extracted by Morgan (Wolf) method

Rating		N	P	K	Ca	Mg	S	B
Poor	*		0-4	0-24	0-24	0-24		0-0.2
		0-14					0-4	
	**		0-9	0-49	0-499	0-49		0.0-0.4
Fair	*		5-15	25-49	250-499	25-49		0.3-0.4
		15-24					5-1	
	**		10-24	50-99	500-999	50-99		0.5-0.9
Good	**		16-49	50-175	500-2499	50-499		0.5-0.9
		25-74					15-199	
			25-62	100-199	1000-4999	100-999		1.0-1.9
High	*	75-200	50-75	176-200	2500-5000	500-1000	200-250	1.0-2.0
	**	75-300	63-87	200-300	5000-10000	1000-2000	200-300	2.0-4.0

*These values for coarse soils of low exchange capacity.
**Suitable values for heavier soils with good CEC values (loam, silt loam, clay loam, and organic soils).

nature. The evaluations are made over a period of time in order to provide average climatic conditions. As a result of the many trials, recommendations for various crops grown on certain soils and with certain climates have been developed. Such data, used by most laboratories for making recommendations, can be made available to growers or other interested parties who want to do their own testing by contacting the public testing laboratories in their respective states. A computer program based on Morgan (Wolf) methods providing recommendations for crops grown on several different soil types is available through MMI Micro-Macro International, Inc., 183 Paradise Blvd., Suite 108, Athens, GA 30607.

It is wise to determine how the recommendations based on soil tests were programmed. The Morgan (Wolf) recommendations are based on high yield goals attained through the use of ample water. Recommended rates of fertilizer vary for different soils to compensate for fixation of several elements or release of N in the different

TABLE 3.12. Average nutrient uptake by good crop yields

Crop	Yield per acre		Primary nutrients			Secondary nutrients				Micronutrients				
		N	P$_2$O$_5$	K$_2$O	Ca	Mg	S	Cl	B	Cu	Fe	Mn	Zn
								lb/ac						
Alfalfa	10 ton	600*	120	600	248	53	51	70	1.60	0.30	1.50	1.20	1.00
Apples	250 cwt	100	46	180	–	24	14	–	0.09	0.08	0.31	0.70	0.23
Barley	80 bu	100	40	80	18	8	–	7	–	–	–	–	–
Beans (snap)	4 ton	138	33	163	–	17	40	–	–	–	–	–	–
Beets (sugar)	30 ton	255	40	530	90	80	41	–	–	0.14	1.10	0.22	–
Beets (table)	500 cwt	360	43	580	–	101	40	–	0.03	0.03	0.20	0.20	0.10
Bermuda (coastal)	10 ton	500	140	420	66	50	26	–	–	–	–	–	–
Birdsfoot trefoil	4 ton	192*	84	272	–	32	20	–	–	–	–	–	–
Bromegrass	5 ton	220	65	315	42	10	64	–	–	–	–	–	–
Cabbage	700 cwt	270	63	249	35	36	–	49	0.15	0.07	0.35	0.18	0.14
Cantalopes	175 cwt	65	21	117	–	12	90	–	–	–	–	–	–
Celery	75 ton	280	165	750	150	45	30	–	0.12	0.12	0.60	0.25	0.12
Clover-grass	6 ton	300*	90	360	–	30	33	–	–	–	–	–	–
Corn (grain)	200 bu	266	114	266	58	65	–	4	–	–	–	–	–
Corn (silage)	32 ton	266	114	266	65	33	11	–	0.23	0.15	1.40	2.10	0.60
Corn (sweet)	90 cwt	140	47	136	–	20	39	–	–	–	–	–	–
Cotton	3 bale	205	90	145	90	36	20	8	0.05	0.23	0.18	0.39	–
Fescue	3.5 ton	135	65	185	21	13	6	–	–	–	–	–	–
Flax	20 bu	54	22	45	–	–	26	–	–	–	–	–	–
Grapes	12 ton	102	35	156	36	18	–	–	–	–	–	–	–
Lettuce	450 cwt	90	30	185	–	9	19	–	0.13	–	1.30	0.34	–
Oats	100 bu	115	40	145	13	20	37	2	0.08	0.08	0.95	0.30	–
Onions	600 cwt	180	80	150	47	18	28	–	–	–	–	–	0.39
Oranges	540 cwt	265	55	330	76	38	–	–	–	–	–	–	–

Crop	Yield per acre		Primary nutrients			Secondary nutrients			Micronutrients					
		N	P$_2$O$_5$	K$_2$O	Ca	Mg	S	Cl	B	Cu	Fe	Mn	Zn
							lb/ac						
Orchardgrass	6 ton	300	100	375	–	25	35	–	–	–	–	–	–
Pangola	11.8 ton	299	108	430	78	67	46	–	0.14	0.14	1.10	1.60	0.49
Peanuts	4000 lb	115	40	145	125	20	19	3	0.08	0.08	0.67	0.53	–
Peas (English)	25 cwt	164	35	105	175	18	10	8	0.04	0.06	0.60	0.40	0.02
Pensacola-bahia	7 ton	303	87	242	–	35	27	–	–	–	–	–	–
Potatoes (Irish)	500 cwt	269	90	546	20	50	22	–	–	0.09	0.36	0.13	0.27
Potatoes (sweet)	300 cwt	156	69	313	25	18	–	–	–	–	–	–	–
Rice	7000 lb	112	60	148	–	14	12	–	–	–	–	–	–
Ryegrass	5 ton	215	85	240	–	40	–	–	–	0.04	–	–	–
Sorghum (forage)	8 ton	198	67	286	–	35	18	–	–	–	–	–	–
Sorghum (grain)	4 ton	250	90	200	65	44	38	–	–	–	–	–	–
Sorghum (Sudan)	8 ton	319	122	467	64	47	32	–	–	–	–	–	–
Soybeans	60 bu	324*	64	142	85	27	25	21	0.09	0.06	1.50	0.56	0.35
Sugarcane	100 ton	360	156	610	120	100	86	–	–	–	–	–	–
Tobacco (burley)	4000 lb	290	37	321	–	33	24	–	–	–	–	–	–
Tobacco (flue-cured)	3000 lb	126	26	257	105	24	19	–	–	–	–	–	–
Tomatoes	30 ton	180	48	336	–	28	41	–	–	–	–	–	–
Wheat	80 bu	134	54	162	21	24	20	35	0.11	0.09	3.20	0.48	0.39

*Legumes normally obtain most of their nitrogen from Rhizobia.

Source: Wolf, B., J. Fleming, and J. Batchelor. 1985. *Fluid Fertilizer Manual.* National Fertilizer Solutions Associations, Manchester, ND 63011. Reprinted by permission of the Agricultural Retailers Association, St. Louis, MO 63146, present publishers of the *Fluid Fertilizer Manual.*

soils. The presence of added OM in the form of manures or cover crops, or lowered requirements because of lower yield goals or restricted water availability, would require adjusting fertilizer use below the recommended levels. Recommended rates using other systems of extraction may also need adjustment, if the growing conditions under which rates were established are markedly different from those faced by the grower. Laboratories will often make these adjustments if this data is provided by the grower, making close cooperation between grower and laboratory essential for the intelligent use of soil tests.

If growers are running the tests on their own, or if the laboratory cannot make the necessary adjustments, amounts of fertilizers to be added or subtracted from a given set of recommendations for varying conditions may be available from the previous field trials used to set up the tests. Again, this data may be obtained from the state agency running soil tests. Often, such specific information is not available but can be estimated. Some of the factors which may necessitate changes in general recommendations are discussed briefly below.

Crop

As pointed out above, the considerable variation in crop requirements is met in some analytical systems by varying the rating of the soil test. In the Morgan (Wolf) computer program, the various crops are rated as to their nutrient requirements and the fertilizer recommendations compensate for these differences. An example of a state testing service using the difference in crop requirements as one of the factors (along with moisture) in making fertilizer recommendations is given in Table 3.13.

Yield Goals

Some typical examples of laboratories using intended crop yields as a basis for making fertilizer recommendations are presented in Tables 3.14, 3.15, 3.16, and 3.17.

If fertilizer recommendations are not based on specific yield goals, but rather on averages obtained over a number of years, some

TABLE 3.13. Recommended phosphate (P_2O_5) and potash (K_2O) applications for several crops grown in Kansas based on soil test and available water

Crop	Irrigated	P soil test level (pp2m or 1 lb/a)				
		VL 0-10	L 11-25	M 26-50	H 51-100	VH 100+
		— — — — lb P_2O_5/a — — — — — — — — —				
Corn	+	60-80	40-60	20-40	0-20	0
	−	40-60	20-40	0-20	0	0
Sorghum	+	50-70	30-50	20-30	0-20	0
	−	40-50	20-40	0-20	0	0
Wheat	+	50-60	30-50	10-30	0	0
	−	40-60	20-40	10-30	0	0
Grasses	+	50-60	40-50	20-40	0-20	0
	−*	30-50	20-40	0-30	0	0
Legumes	+	90-100	60-90	30-60	0-30	0
	−	40-60	30-40	20-30	0-20	0
		K soil test level (pp2m or 1lb/a)				
		VL 0-80	L 81-160	M 161-240	H 241-320	VH 320+
		— — — — lb K_2O/a — — — — — — — — — —				
Corn	+	80-100	60-80	40-60	20-40	0
	−	60-80	40-60	20-40	0-20	0
Sorghum	+	80-100	60-80	40-60	20-40	0
	−*	50-70	30-50	10-30	0	0
	−**	60-80	40-60	20-40	0-20	0
Wheat	*	40-60	20-40	0-20	0	0
	**	20-30	10-20	0-10	0	0
Grasses	+	50-60	40-50	20-40	0-20	0
	−	40-50	30-40	20-30	0-20	0
Legumes	+	80-100	60-80	40-60	20-40	0
	−	80-100	50-80	30-50	0-30	0

*Moderate rainfall region.
**Poor rainfall region.
Wheat recommendations are for nonirrigated farms but with different rainfall.

Source: Whitney, D. A. 1983. Soil interpretations and fertilizer recommendations. C-509, Coop. Ext. Service, Kansas State Univ., Manhattan, KS.

adjustments in fertilizer recommendations may need to be made if intended yield goals are much above or below established averages. Table 3.18 lists the amounts of added nutrients needed for incremental increases of several crop yields, and can be used as a basis for increasing or decreasing fertilizer recommendations to meet these lower or higher yield goals.

Soil Variation

Different fertilizer recommendations for dissimilar soils, despite the same soil tests, are made to compensate for variations in: (1) fixation of P and/or K; (2) release of N related to OM differences; and (3) leaching losses of N, as well as K and Mg.

The fixation of P and K is related to the amount and kind of clay in the soil. Usually, different recommendations are made for soils of different textural classes or CEC, with the largest amounts of P (and sometimes K) being applied to the heavier soils and lesser amounts to the sands.

Some idea of the different fertilizer requirements of soils with varying clay content can be gained from the following: (1) About four times as much K is needed to maintain a desirable level of K in a heavy soil with 25 percent clay as compared to a light soil with only 3 percent clay; (2) Sandy and clay loams require an application of 4 to 5 lbs P per acre to change the Mehlich I test 1 lb/a while about 12 lbs are required to produce a similar change in a clay soil. For example, adding 50 lbs P/a to a clay soil that tests 10 lbs P/a only raises the available P test to about 14 lbs P/a while adding the same amount to a sandy loam that also originally tested 10 lbs P/a could be expected to raise the P test to about 20 lbs/a.

Since the extra P and K needed for the heavier soils is a result of fixation, banding of fertilizer can greatly limit the need for more P and K of these soils. It has been estimated that the amount of P required when applied in bands to soils high in clay (silt loams, clay loams, and clays) is about 1/3 of that required if applied broadcast. Such differences in the effectiveness of P and K applied in bands as compared to broadcast are reduced as the sand content is increased (i.e., clay is decreased). The computer program for the Morgan (Wolf) methods is based on broadcast applications of P and K, making it desirable to reduce the recommendations for these ele-

TABLE 3.14. Phosphate and potash recommendations for small grains grown in Indiana based on soil tests and yield goals

Soil	P test*	Crops	Yield goals in bu/a				
		Wheat or rye	30-44	45-54	55-64	65-74	75+
Lbs/a	Rating	Barley or oats	70-85	86-100	101-115	116-130	130+
			— — — — — — lbs/a P_2O_5 — — — — —				
0-10	VL		90	120	120	140	140
11-20	L		60	90	90	110	110
21-30	M		30	60	60	90	90
31-50	H		20	30	30	60	60
51+	VH		20	20	20	20	20
To maintain H or VH test			20	30	40	50	60
Soil K test**							
Lbs/a	Rating		— — — — — lbs/K_2O — — — — — —				
0-80	VL		90	120	120	120	120
81-150	L		60	90	90	90	90
151-210	M		30	60	60	60	60
211-300	H		0	30	30	30	30
301+	VH		0	0	0	0	0
To maintain H or VH test			15	20	20	25	30

*Bray-Kurtz #1 method **Ammonium acetate, 7.0

TABLE 3.15. Nitrogen recommendations for different corn yield goals grown with and without manure or legumes

Legume or manure	Yield goal per acre			
	60-89 bu	90-110 bu	120-149 bu	150-180 bu
Legume, 10 tons manure/a	0	0	50	100
Good legume	10	40	90	140
10 tons manure/a	30	60	110	160
No legume, no manure	70	100	150	200

Source: Warnecke, D. D., D. R. Christenson, and R. E. Lucas. 1976. *Fertilizer Recommendations for Vegetables and Field Crops.* Ext. Bull. E-550. Farm Sci. Series, Coop. Ext. Serv., Michigan State Univ., E. Lansing, MI.

TABLE 3.16. Nitrogen recommendations for several crops grown in Michigan based on legume, with and without manure addition

Previous crop or manure application	Barley Oats Rye	Wheat	Field beans Soybeans	Grass Management low	high	Potatoes–cwt/a* 250-349	350-449	450-550
Legume + 10 tons manure/a	10	10	0	0	0	30	60	90
Good legume	10	10	10	0	0	70	100	130
10 manure/a	10	30	10	0	50	90	120	150
No legume or manure	40	60	40	60	100	130	160	190

*Yield goals
Source: Warnecke, D. D., D. R. Christenson, and R. E. Lucas. 1976. *Fertilizer Recommendations for Vegetables and Field Crops.* Ext. Bull. E-550. Farm Sci. Series, Coop. Ext. Serv., Michigan State Univ., E. Lansing, MI.

TABLE 3.17. Indiana nitrogen recommendations for corn or grain sorghum based on yield goals and previous crop

Previous crop	Yield goals in bu/a				
	100-110	111-125	126-150	151-175	176-200
	— — — — — N lbs/a — — — —
Good legume (5 plants)/square feet (alfalfa, red clover)	40	70	100	120	150
Average legume (2-4 plants) (alfalfa, red clover)	60	100	140	160	180
Soybeans. Seeding of alfalfa, red clover	100	120	168	190	220
Corn, small grain or grass crop	120	140	170	200	230

Source: Plant and Soil Test Laboratory, Agronomy Dept., Purdue Univ., Lafayette, IN. Courtesy of Prof. R. K. Stivers.

TABLE 3.18. Amount of nutrients* needed for added increments of crop yields

Crop	Unit	N	P₂O₅	K₂O	S	Ca	Mg	Zn	Fe	Mn	Cu	B
							lb/a					
Alfalfa	ton	60.00**	73.60	82.80	21.00	56.00	13.00	0.32	0.39	0.28	0.038	0.08
Barley	bu	1.40	3.90	1.80	0.75	0.43	0.23	0.009	0.01	0.02	0.002	0.001
Beans	bu	3.50	6.40	1.60	0.88	0.28	0.04	0.006	0.03	0.003	0.002	0.003
Bermuda (16% protein)	ton	57.00	45.80	62.40	11.00	13.00	8.00	0.18	0.24	0.50	0.07	0.09
Clover-grass	ton	52.00**	48.50	60.00	11.00	42.00	10.00	0.16	0.30	0.20	0.05	0.04
Clover (various)	ton	30.00***	112.20	87.60	10.00	50.00	11.00	0.17	0.26	0.14	0.05	0.05
Corn (grain)	bu	1.80	1.20	1.40	0.30	0.25	0.27	0.045	0.05	0.03	0.002	0.008
Corn (silage)	ton	52.00	50.40	43.20	10.00	12.00	16.00	0.30	0.50	0.25	0.04	0.06
Cotton (lint)	bale	62.00	178.60	80.40	17.00	34.00	12.00	0.32	0.46	0.16	0.06	0.50
Grass (various)	ton	57.00	52.70	68.40	13.00	12.00	6.00	0.14	0.36	0.28	0.05	0.02
Ladino clover	ton	93.00**	103.10	129.60	12.00	58.00	20.00	0.28	0.96	0.42	0.03	0.03
Oats	bu	1.20	2.70	1.10	0.17	0.25	0.33	0.008	0.01	0.004	0.001	0.001
Peanuts	cwt	1.20**	10.50	6.70	1.46	3.65	0.19	0.006	0.03	0.01	0.001	0.02
Potatoes (Irish)	bu	3.20	6.60	6.20	0.30	0.22	0.70	0.002	0.12	0.003	0.002	0.001
Potatoes (sweet)	bu	2.40	4.40	4.80	0.40	0.28	1.00	0.002	0.13	0.004	0.002	0.002
Rice	bbl	3.40	6.40	3.20	0.80	0.60	0.90	0.006	–	0.03	0.007	–
Rice	bu	9.40	17.60	9.00	2.20	1.70	2.50	0.017	–	0.08	0.019	–
Sorghum (milo)	cwt	3.00	1.40	1.40	0.33	0.30	0.16	0.01	0.01	0.03	0.006	0.001
Sorghum (silage)	ton	46.00	75.60	48.00	8.00	6.00	3.00	0.01	0.12	0.01	0.003	0.01
Soybeans	bu	1.40**	6.60	3.80	0.63	2.30	1.50	0.014	0.13	0.03	0.005	0.006
Sudan	ton	50.00	52.70	48.00	8.00	15.00	19.00	0.10	0.40	0.20	0.03	0.02
Sugar cane	ton	3.60	13.50	16.60	3.40	1.80	2.00	–	–	–	–	–
Sugar beets	ton	4.50	11.00	6.10	4.00	4.30	3.20	0.15	0.03	0.10	0.004	0.001
Tomatoes	ton	8.40	15.60	13.40	1.30	0.30	0.50	0.02	0.10	0.01	0.009	0.01
Wheat	bu	1.80	5.50	1.70	0.36	0.32	0.56	0.007	0.02	0.06	0.002	0.001

* Expected efficiency of nutrient added as fertilizers from the most common sources.
** Legumes obtain most of their nitrogen from Rhizobia.
Source: Wolf, B., J. Fleming, and J. Batchelor. 1985. *Fluid Fertilizer Manual.* National Fertilizer Solutions Association, Manchester, ND 63011. Reprinted by permission of the Agricultural Retailers Association, St. Louis, MO 63146, present publishers of the *Fluid Fertilizer Manual.*

ments if the fertilizer is to be applied in bands. We have used the following percent reductions in recommended P or K if applied in bands for the different soil types: 60 percent for silt and clay loams; 50 percent for loams; 30 percent for sandy and gravelly loams; and 25 percent for sands. In most cases, the reduced amounts of banded fertilizer have produced yields equal to the larger broadcast amounts.

In a few cases, yields have been higher with the larger broadcast amounts probably due to one of the following: (1) Soils contained large amounts of OM which limits the fixation of P and K; (2) Soils contained clay with relatively low fixation powers due to the type of clay or partial saturation of the clay with previous large applications of these elements and (3) Bands were not positioned properly for maximum uptake by the plant. In a few cases, higher yields are obtained with the smaller amounts of banded fertilizer, probably due to better positioning of nutrients in cold or dry soils.

In cold soils, nutrient uptake is greater if nutrients are positioned close to developing roots. In dry soils, uptake can be improved if fertilizer is placed deeper in the soil where moisture is more plentiful.

The depth of the soil is an important factor in determining fertilizer requirements because of the extent of an effective root system. Shallow soils, because of moisture (shortage or excess), subsoil pH, compaction, or hardpans, require more fertilizer, but may need to have the application divided in order to avoid producing damaging concentrations of salts.

Fertilizer recommendations also need to take into account differences between soils that are caused by leaching. The problem of N loss, and at times K and Mg as well, is of primary concern for light soils in regions with heavy rainfall. Rainfall distribution may be more important than total rainfall. For example, a single rain of about 1 inch falling in a few hours on several Florida sandy soils (CEC less than 5 meq/100 g) with ideal moisture contents will completely deplete the N and much of the K. Losses of Mg, SO_4 and even P will also be considerable. The same quantity of rain falling over several days will have little impact. Although not frequently encountered, the problem of leaching can also be serious in arid climates if irrigation application is much greater than the soil's capacity to hold water.

Some of these losses can be avoided by choosing less soluble forms of the nutrients: dolomitic limestone rather than soluble Mg; superphosphates instead of the ammonium phosphates; and SCU, Uramite, or Osmocote instead of ammonium, nitrate, or urea as sources of N.

Another method of avoiding nutrient losses is to time the application of the fertilizer with sidedressings. It is highly practical to save some of the fertilizer—or at least the N—for later application in regions where leaching losses can be expected, and use later soil test results to determine the amounts of fertilizer to apply.

Available Water

Soil moisture is a major factor affecting nutrient release and plant growth. The amounts of nutrients that are adequate with low moisture may be inadequate with ideal moisture. Some states with regions that receive different amounts of rainfall, or have irrigated and nonirrigated crops, may also have different fertilizer recommendations for the different water regimes. Some variations of P_2O_5 and K_2O applications for several crops grown in Kansas on irrigated and nonirrigated soils having different soil test readings are given in Table 3.13 (p. 87).

Organic Matter Additions and Nitrogen

Because soil OM has an effect on the amounts of N available for the crop, several systems use OM as a basis for recommending N fertilizer, as exemplified by the Arkansas recommendations for cottons. (See Tables 3.19 and 3.20.)

In the northern states, serious N deficiencies can arise in the spring when soils are too cold for N release. Eliminating or greatly reducing the N application because of large amounts of soil OM can seriously limit crop yields. A better approach for such areas, is to measure the available NO_3 and NH_4-N and eliminate or greatly reduce the N application only if these are present at high levels. If soils contain good to high levels of OM, or they have received manures, cover crops, or other organic matter additions, only about 1/3 of the recommended N application should be added at planting.

TABLE 3.19. Nitrogen recommendations for cotton as affected by soil organic matter

Soil OM	Northern Arkansas	Southern Arkansas
%	lb/a	lb/a
3+	40-50	45-50
2-3	50-55	55-60
1-2	60-65	60-65
<1	65-70	70-80

Source: Miley, W. N. *Fertilizing Cotton with Nitrogen.* Leaflet 526. Univ. of Arkansas Coop. Ext. Serv.

TABLE 3.20. Reductions in recommended amounts of nitrogen for cotton grown in different sections of Arkansas on soils with different organic matter contents

Nitrate-N	Northern Arkansas		Southern Arkansas	
in	% soil OM		% soil OM	
upper 18 in.	0-1.9	2.0+	0-1.9	2.0+
lb/a	Reduce N-lb/a		Reduce N-lb/a	
0-15	0	0	0	0
15-30	10-15	10-20	0	0
30-45	20-25	25-30	10-15	20-25
45-60	25-30	30-40	20-25	30-35
60+	35-45	40-50	30-35	35-40

Source: Miley, W. N. *Fertilizing Cotton with Nitrogen.* Leaflet 526. Univ. of Arkansas Coop. Ext. Serv.

In the Morgan (Wolf) method, the amount added is enough to bring the available soil N from a Poor or Fair level to a Good level. (See Table 3.11 for amounts of nutrients held at different ratings.) For example, if only 10 lbs./a of available N were present and no manure or legumes were added, it would be desirable to raise the available N to midway in the Good range, and so 40 ppm of N (or 80 lbs/a) would be needed. The addition of manures or legumes as potential sources of relatively available N permits using the much lower nutrient levels of the Fair range (25-49 ppm) as a guide. In this case an application of only 17 ppm (34 lb/a) is enough to bring available N into this range. This method will work only if later tests are made to determine whether sidedressings of N are needed.

Use of tests for available N can be helpful even if tests are made only prior to planting the crop. Kansas suggests the following reductions in N rates based on varying levels of NO_3-N + NH_4-N in the first two feet of soil:

N Rating	*Adjustment in Fertilizer Recommendation*
Low	No reduction
Medium	Reduce by 1/3
High	Reduce by 2/3
Very high	No N recommended

Manures can supply appreciable amounts of N and several other elements, but the release of the nutrients are also affected by soil factors as cited for OM. Only about 1/2 of the N can be expected to be released for the first crop year. Although manure contains 15-20 pounds of N per ton, the University of Maryland assumes only about a 5-pound N release per ton, as it suggests a 50-lb/a reduction in N application for corn receiving 10 tons/a of manure as compared to no manure. An assumption of 33-50 percent release in the first year is realistic, but obtaining maximum economic yield (MEY) without causing nitrate pollution of ground waters will require reduced N fertilization at planting, subsequent measurement of available N, and applications of N sidedressing as tests indicate the need.

Other OM sources also affect the amounts of available N. Applications of various animal waste or good leguminous crops can contribute much of the N needed by the succeeding crop, but as with OM and manures, the major release of these nutrients is affected by microorganism activity and only about 1/3 to 1/2 can be expected in the first crop year.

Nonleguminous cover crops will provide little N or may even rob N from the succeeding crop. Mature nonlegumes have such wide C:N ratios that the decomposition of the material by microorganisms tends to tie up available N. No initial release of N can be expected if the cover crop is less than 1.75 percent N; extra N must be added at planting time if the N content of the cover crop (or any other OM) is less than 1.3 percent.

Relatively immature nonlegumes can be expected to release N and several other elements for the succeeding crop, especially if they followed a cash crop which had been fertilized heavily. The amounts of nutrients released by such nonlegume crops are difficult to estimate. The nutrient value per acre of such crops can be determined by multiplying the average dry weight of a square foot of cover crop by percent of the different nutrients by 43,560. A release of 33-50 percent of the total nutrients can be expected in the first year.

The amounts of N fertilizer that will be needed in the presence of applications of manure or previous leguminous crop will depend on several factors, such as crop to be grown, yield goals for the crop, and the kind and quality of the cover crop. Some N recommendations for several crops grown under these situations are given in Tables 3.16, 3.17, and 3.20.

The Vegetable Research Trust in Warwick, England has published an N (along with a P and K) Predictor* that uses OM to provide an N Index. Nitrogen response values established for different amounts of N applications at different indices for 20 vegetables allows the grower to choose a level of fertilization capable of giving maximum yields or any acceptable portion thereof. Four N indices are established which correspond to the following.

0. Very low OM soils previously cropped for at least two years with cereals.
1. Normal soil cropped one year with cereals or storage root crops.
2. Normal soil in arable or ley (fallow) systems.
3. Soils with large N reserves, or that have received heavy applications of farmyard manure or other organic manures.

Recommended applications of N (in kg/hectare**) to obtain maximum yields of a highly responsive crop such as table beets on the different indices are 120 for #1; 90 for #2; and 70 for #3. For a

---

*The NPK Predictor can be purchased by contacting The Liason Officer NVRS, Wellesbourne, Warwick CV35 9EF, England.

**Multiply by 0.893 to obtain lbs/a.

less-responsive crop such as broad beans, it is 20 for #0; 15 for #1; 10 for #2; and 10 for #3.

High Nutrient Levels and Fertilizer Recommendations

In recent years, our own experience and a considerable body of evidence has accumulated indicating that small applications (starter amounts) of both P and K applied at planting are helpful for many crops even if tests for these elements appear to be very good. Evidently, cold or compact soils slow root growth, so that the addition of quickly available nutrients very close to the emerging root enables the plant to make a quick start. Because of these responses, we have recommended very small amounts of P and K_2O strategically placed close to the seed, even if tests for these elements indicate that levels are adequate. If the N test is not high, a small amount of N is included with the fertilizer. Amounts have been limited to about 10 lbs/a of N + K_2O placed with the seed. No recommendations for added N or K_2O are made if conductivity (soluble salts) is in the high range, but the addition of P as superphosphates is permitted

Additional Reading

Aljibury, F. K., J. L. Meyer, and W. E. Wildman. 1982. *Managing Compact and Layered Soils.* Leaflet 2635. Cooperative Extension, Division of Agricultural Science, University of California, Berkeley, CA 94720.

Brown, J. R. (ed.). 1987. *Soil Testing: Sampling, Correlation, Calibration, and Interpretation.* SSSA Special Publication #21, Soil Science Society of America, Inc., Madison, WI 53711.

Chancellor, W. J. 1977. *Compaction of Soil by Agricultural Equipment.* Bull. 1881. Cooperative Extension Division of Agricultural Science, University of California, Berkeley, CA 94720.

Glinski, J. and J. Lipiec. 1990. *Soil Physical Problems.* CRC Press Inc., Boca Raton, FL.

Harpstead, M. I. and F. D. Hole. *Soil Science Simplified.* 1981. Iowa State University Press, Ames, IA 50010.

Hillel, D. 1982. *Introduction to Soil Physics.* Academic Press, New York, London.

Jones, J. B., Jr. 1981. *Soil Testing Handbook.* Benton Laboratories, Athens, GA 30604.

Letey, J. 1985. Relationship between soil physical properties and crop production. pp 277-294. In *Advances in Soil Science*, Vol. 1, Springer-Verlag, New York.

Naderman, G. C. *Subsurface Compaction and Subsoiling in North Carolina*, AG-353. Extension Service. North Carolina State University, Raleigh, NC.

Neja, R. A., W. E. Wildman, and I. P. Christensen. 1982. *How to Appraise Soil Physical Factors for Irrigated Vineyards.* Leaflet 2946. Cooperative Extension, Division of Agricultural Science, University of California, Berkeley, CA 94720.

Wells, K. L. and W. R. Thompson (eds.). 1992. *Current Viewpoints on the Use of Soil Nitrate Tests in the South.* ASA Misc. Publication, American Society of Agronomy, Inc., Madison, WI 53711.

Westerman, R. L. (ed.). 1990. *Soil Testing and Plant Analysis.* Soil Science Society of America, Inc., Madison, WI 53711.

Wildman, W. E. 1982. *Diagnosing Soil Physical Problems.* Leaflet 2664. Cooperative Extension Service, Division of Agricultural Sciences, University of California, Berkeley, CA 94720.

SECTION II:
PLANT DIAGNOSTICS

There are at least three areas in which plant diagnostics are useful in increasing crop production: (1) assessing purity, germination, and vigor of seeds; (2) evaluating the nutritional status of the plant; and (3) determining maturity or the ideal time of harvest.

Chapter 4

Assessing Potential Performance of Seeds

Crop yields are closely correlated with the purity, germination, and vigor of seeds. Purity and germination are regulated by various state and federal laws, and their evaluations are often provided on tags accompanying the seed lot. The extent of germination after some storage as well as the vigor of the seed lot can be assessed by special tests.

PURITY

Purity of a seed lot is measured by the absence of foreign matter. The presence of other crop seeds, weed seeds, and any foreign debris detract from purity. The presence of these foreign materials can be physically measured and may be noted on the seed tag.

Cultivar purity is also of concern to the grower, especially with the high performance and cost of so many new hybrids. Some cultivars can be identified by special physical traits, such as seed and/or hilum color, but often these characteristics are not positive. In such cases, the cultivars may be identified by protein, enzyme, or DNA analysis. Such analyses, however, are highly technical, requiring considerable equipment and expertise, and making such tests very expensive. Simplified laboratory electrophoresis tests for cultivar identification and genetic purity assays of brassicas, carrots, corn, curcurbits, forage and lawn grasses, lettuce, oil crops, onions, peas, potatoes, small grains, and tomatoes are available from Isolab, Inc., Akron, OH 44321. But in most cases, the grower is dependent upon the integrity of the seed supplier to obtain true seed cultivars.

GERMINATION TESTS

Germination tests are required by many states and appear on most lot tags. It is often desirable to repeat the test in order to evaluate current germination after some storage. An approximate evaluation can be obtained by placing seeds between moistened blotters or filter papers, although larger seeds, such as peas or beans, can be placed in moistened sand. Watering the seeds initially with 0.2 percent potassium nitrate solution overcomes dormancy. The seeds are then maintained with sufficient water to replace that which is lost, but excesses should be avoided as insufficient oxygen will interfere with germination. Generally, blotters or filter papers should not be so wet that films of water form around seeds or if pressed a film of water forms around the finger. Seeds are allowed to germinate in a chamber or closed area where about 95 percent humidity can be maintained.

Many seeds can be germinated at about 86°F, but some, such as celery, lettuce, parsnip, English pea, and spinach, will do better in the 60-70°F range. The temperature for celery needs to be fluctuated from a high of about 70°F during the day to less than 60°F at night. All seeds need to be examined after several days, at which time most cantaloupe, sweet corn, cauliflower, cucumber, radish, turnip, and watermelon will have germinated. An inspection at seven days and again at 14 days will be needed for slower germinating seeds.

Counts of normal viable plants and abnormal seedlings are made at each examination. Seedlings are examined for freedom from defects of roots, hypocotyl, and epicotyl. Weak seedlings are those that are missing primary roots or a cotyledon; have breaks, lesions, necrosis, or abnormal hypocotyls; or show signs of decay of the primary root or hypocotyl.

If precise germination data are required, the evaluation must be conducted under the very exacting conditions set by Official Seed Analysts. A current copy of official rules for testing seeds can be ordered through Larry J. Prentice, Association of Official Seed Analysts, Inc., 268 Plant Science, IANR-UNL, Lincoln, NE 68503-0911.

VIGOR TESTS

Usual germination tests are helpful, but do not provide information about seed performance under adverse conditions. For such information, the grower may resort to one or more of the tests listed below.

Accelerated Aging

This stress test, originally designed to predict soybean seed storage life, is being used to test the vigor of soybeans, cereal grains, peanuts, and many vegetables. Germination is noted after seeds are incubated for a few weeks at high relative humidity (100 percent) and temperatures of 104-113°F.

Cold Testing

Another stress test attempts to predict the germination of seeds planted in wet, cold soils. It has become widely used to measure the vigor of corn seeds. Seeds are planted in cold, nonsterilized soil at 70 percent moisture holding capacity (MHC) or placed on paper towels lined with soil, and kept for seven days at 50°F, followed by four days at 77°F. A high rate of germination at the end of the period indicates a vigorous seed able to tolerate soil pathogens in a cold, wet soil.

The tolerance of seeds to low temperatures can be determined by germinating seeds on blotters at a constant temperature of 65°F instead of 86°F and making the counts of germinated seeds in six to seven days. This type of test has been especially useful for evaluating cotton seed.

A modified cold test uses 50 seeds on a wet towel placed on wax paper larger than the towel. The seeds are covered with another wet towel, loosely rolled together and held in place with rubber bands. A total of 200 seeds are used for the test. The seed rolls are placed in a germinator at 50°F for five days, at which time they are removed and placed in a germinator at 68-86°F for an additional five to seven days. At the end of the test, counts of germinating seeds and defects are compared with 300 seeds subjected to the standard germination test.

Conductivity Tests

Soaking 50 seeds in 250 ml of water for 24 hours at 68°F and measuring the conductivity of the filtered water indicates the loss of soluble sugars, starches, and proteins. The conductivity measures the amount of leakage and is an indicator of membrane and seed coat integrity.

Dry Weight

Seedling vigor is indicated by the dry weight of seedlings separated from cotyledons after seeds have been germinated on paper towels in a darkened germinator for seven days.

Root Elongation

Measurement of developing roots in the standard germination tests with seeds oriented so that the radicle elongates downward can also be used to denote vigor of some seeds, such as lettuce.

Tetrazolium Chloride

The intensity of color produced by staining the seed embryo with triphenyl tetrazolium indicates the respiratory rate and indirectly the viability and vigor of the seed. Rules for conducting the tetrazolium test can also be obtained by contacting Larry J. Prentice, Association of Official Seed Analysts, Inc., 268 Plant Science, IANR-UNL, Lincoln, NE 68503-0911.

Chapter 5

Plant Nutrition

The usefulness of soil testing in determining nutrient needs can be greatly improved by combining it with plant analysis (sometimes known as leaf analysis) whenever possible. Plant analysis supplies information as to the nutrients taken up by the plant and at times can be a better indicator of nutrient availability than the soil tests. Plant analysis is used to: (1) verify plant deficiency or toxicity symptoms; (2) determine the adequacy of the fertilizer programs; and (3) provide a basis for recommending fertilizer.

Two different types of plant analyses are commonly used. One type, known as tissue tests, determines the amounts of soluble or readily soluble nutrients in the sap or in macerated fresh tissue. It is used primarily for the determination of N, P, and K, but can also be used for Ca and Mg. Generally, tissue tests lack the sensitivity to determine the micronutrients. The second procedure, which is more commonly used, determines the total amount of nutrients. It requires drying of the plant sample and destruction of the OM, but permits determination of the micronutrients as well as the macronutrients. The latter method requires much more time, equipment, and know-how, but yields more complete and quantitative results.

The average grower will be better served by using established laboratories to conduct total analyses, but some types of tissue testing can also be used to advantage. Kits for such purposes are available and some of these are listed in Appendix 5.

SAMPLING

The Importance of Correct Sampling

The utilization of a leaf analysis is dependent upon precise sampling. The distribution of elements within the plant is markedly different in the different plant organs, and even within different parts of the various organs. Nutrient distribution in the leaf blade will usually be far different than in the petiole, and different parts of the leaf will have vastly different nutrient concentrations. Other factors which affect plant composition are position of the leaf, age of tissue, shade, and, for NO_3-N, even the time of day. Because of all these variables, plant sampling must follow very rigid rules governed by conditions under which the sufficiency values or norms were established.

Selection of Plants

Because stress can also affect plant composition, plants affected by disease, insect attack, mechanical injury, spray damage, drought, or excess moisture should not be selected for routine sampling.

There is considerable variation among plants in the field. The selection of plants from which the plant parts are removed will vary to some extent with the purpose of the analysis. For routine evaluation of a fertilizer program, selected plants must represent the norm or average. If the plant analysis is to be used to diagnose the cause of poor performance of individual plants, two separate samples of poor and normal plants will be needed. In selecting the poor plants, take care to avoid plants that may be of different age or those that have been subjected to nonnutritional stresses.

Number of Plants

A number of plants must be selected to limit variation between plants. The number given in sampling directions is usually the minimum needed to avoid undesirable variations between duplicate samples.

Sample Preparation

Handling Samples

Respiration, by consuming carbohydrates, alters the relationship of the nutrients to dry weight. Drying or refrigerating the sample essentially stops the process. Fresh samples for laboratory evaluation must be rushed to the laboratory as soon as possible. Cooling the sample to about 45°F before shipping helps limit respiration, but plant material must be free of surface moisture before placing in bags. Wrapping individual plant parts in dry paper towels or tissues prior to shipping decreases the chance of spoilage. If the sample designated for total analysis will not reach the laboratory within about 48 hours, it must be dried before or during shipping. The sample can be dried by placing it in an oven overnight at 175°F or in open mesh bags prior to shipping. The former method allows for the decontamination of the sample by washing before drying.

Washing the Sample

Dusts and spray residues can collect on the plant. Dust compromises the tests for Al, Fe, and Si. Spray residues may affect analysis for Cu, Fe, Mn, and Zn. Dust is usually not a problem in regions of ample rainfall or where overhead irrigation has been frequently used.

Washing the sample prior to drying can appreciably lessen dust contamination and to some extent that of spray contamination. The sample is first cut up into pieces of 1-2 inches. The pieces are placed in a deep dish or small pan and washed with gentle rubbing in a 1 percent P-free detergent solution. Exposure to the solution is limited and is followed with a rinse of tap water and two rinses of deionized or distilled water. The washed plant parts need to be drained quickly and placed in clean pans or cloth bags in an oven, set at 175°F.

Drying the Sample

Drying in forced draft ovens is preferred, but satisfactory drying can be obtained by placing the pans with the drained parts in a baking oven with the door held ajar.

Although tissue testing is done with fresh material, some systems call for the drying of appropriate samples in order to have a common reference point for expressing the analytical data.

Particle Size Reduction

Total analysis requires the destruction of the OM prior to element determination. The dried plant material needs to be reduced to very small sizes in order to hasten the decomposition. Reduction to small particles also permits better precision in sampling. If the sample size is smaller than 1.0 gram, it is desirable to reduce particle size to that which passes through a 40-mesh screen. Larger sample sizes can use a 20-mesh screen. Reduction in size of dry samples is accomplished using grinding or ball mills. Contamination of the sample is a possibility, but the use of blades made from tool or stainless steel decreases the chances of contamination in grinding mills.

SOLUBILIZING THE NUTRIENTS

Total Analysis

Destruction of OM

In total analysis, the nutrient elements are freed as the OM is destroyed, which is commonly done by dry or wet ashing. Dry ashing is accomplished by raising the temperature in a muffle furnace; wet ashing by the use of heat combined with acids. Both processes require considerable time, expensive equipment, and considerable expertise. Both procedures are outlined in the *Plant Analysis Handbook* listed under "Additional Readings" at the end of Section II.

Tissue Analysis

Extraction of Nutrients

Essentially there are two different methods of measuring nutrients in fresh tissue for tissue analysis.

In "sap analysis," only nutrients present in the sap are analyzed. The sap is pressed out by devices, such as a garlic press, and the nutrients are measured directly in the sap. These methods are very quick and can be run easily in the field.

Although the other methods take longer than the sap testing, they are still quick, and are best run in a laboratory. Fresh minced material is usually macerated for a period of about five minutes in a Waring blender, using a mild extractant such as acetic acid or Morgan's solution for the extraction and activated charcoal as a decolorizing agent. A portion of the fresh sample is dried at 175°F. In some methods, dry tissue is minced and extracted with water or acetic acid prior to filtration. In either case, the solution is filtered prior to determination of the macronutrients. Both nutrients in the sap and those easily extracted from other tissues are determined and the results are calculated on a dried-weight basis and are then compared with sufficiency standards.

METHODS OF ANALYSIS

Total Analysis

Nutrients released by wet or dry ashing are determined by methods and equipment similar to that used for soil testing. Colorimetric or turbidimetric tests can be used for most elements, but these have largely given way to atomic adsorption (with the exceptions of N, S, and F) or inductively coupled plasma emission spectrometers (ICP or ICAP). Kjeldahl digestion is still being used extensively for N determination, and ion electrodes are still used to some extent for determining NO_3-N, NH_4-N, Ca, Cl, Na, and F. Most of these are laboratory procedures requiring considerable expertise and expensive equipment, although ion electrodes are being used to a limited extent in the field. Of the lab procedures, colorimetric or turbidimetric procedures require the least amount of initial outlay.

Tissue Analysis

Nutrients extracted by a garlic press or mincing by tissues are often determined by test papers dipped in the sap. Colors developing in the papers are compared with color plates, color wheels, or standard test papers. Alternatively, the expressed plant sap, sometimes filtered, can be placed on spot plates or in small vials, and as chemicals are added, developing colors or turbidities are compared with color plates or standards produced in the same manner. Ion electrodes are also being used to determine NO_3-N, NH_4-N, Cl, and K. The test papers and ion electrodes are well adapted for field use.

INTERPRETATION OF RESULTS

Evaluation Methods for Total Analysis

Plant analysis data can be used to interpret the nutrient status of crops because of the general relationship between plant nutrients and crop yield that is expressed in Figure 5.1. Two basic methods of interpreting analysis data are now being used.

The more common and historical approach determines whether a particular analysis falls within a range of nutrients considered satisfactory or above a critical or deficient level (d as in Figure 5.1). The critical level has been defined in various ways, but more recently has come to mean the point below which the yield is reduced by 10 percent. This in turn has been enlarged to include an upper critical or toxic level above which yields are depressed by 10 percent.

In recent years, the interpretation of plant analysis has also used a concept introduced by E. R. Beaufils of the University of Natal, South Africa, which compared element ratios found within the plant to norms established from the examination of many plant analyses. This type of interpretation is designated the Diagnosis and Recommended Integrated System (DRIS).

The older system has the advantage of simplicity and a great deal

FIGURE 5.1. Relationship of plant nutrient composition with available soil nutrients

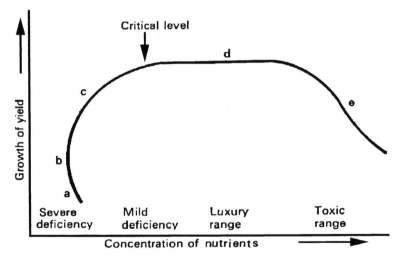

Source: Smith, P. F. 1962. Mineral analysis of plant tissues. *Plant Physiol.* 13: 81-108. Reproduced with permission from the Annual Review of Plant Physiology, Vol. 13, ©1962 by Annual Reviews, Inc.

more historical information. The DRIS system has the advantage of evaluating the various relationships of one element to another and is probably better able to predict the element or elements that might be limiting production. It still has the disadvantage of incomplete norms, especially for the micronutrients in many plants.

Sufficiency Data

Data to evaluate the sufficiency of a plant analysis have been published in many scientific articles and summarized in a number of books. Some of the more important books are listed at the end of this chapter. Ideal ranges of total nutrients for a number of crops are given in Table 5.1 (pp. 116-143), for some common tissue tests in Table 5.2 (pp. 144-148), and for sap analysis in Table 5.3 (pp. 149-151).

DRIS Evaluations

A good description of the DRIS method for evaluating is given in the *Plant Analysis Handbook* listed under "Additional Readings" at the end of Section II.

Plant Test Interpretation

The test results are evaluated by comparison with standards, using critical values to define nutrient status. Data determined with laboratory procedures are compared to sufficiency standards, such as those presented in Tables 5.1 and 5.2. Data from field determinations are usually not quantitative and colors of spot tests or papers are compared with those that represent various stages of adequacy.

USING PLANT ANALYSIS DATA

Precautions

Plant analysis is a tool that can greatly aid the grower, but it must be used with caution. In order to avoid obtaining misleading information, the following minimum precautions need to be observed

1. Be absolutely sure that the sample is collected according to the guidelines as presented for the different plants in Tables 5.1, 5.2,

and 5.3. Failure to do so may make the sample worthless because of insufficient data to evaluate the results or, worse, it will be evaluated improperly. Particular attention must be paid to the collection of complete plant parts as called for in the sampling directions.

2. Collect a corresponding soil sample at the same time as the leaf sample. This will greatly aid in interpreting the leaf analysis. Levels of several nutrients in the plant may be low despite ample levels in the soil because several nonnutrient factors (pH, salts, moisture, and soil compaction) are interfering with normal growth and absorption. Adding nutrients to the soil at such times is nonproductive but foliar applications can be useful.

Repetitive Analyses

By repetition of plant analysis or "crop logging," it is often possible to adjust fertilizer sidedressings and foliar sprays to maximize the crop yield with the lowest possible fertilizer use.

Full use of plant analysis requires early sampling, which may present some problems. There may not be good standards to evaluate very early samples. Attempts are being made to establish sufficiency levels for younger plants. This has been done for the total analysis of cantaloupes, sweet corn, and trellis tomatoes, and effectively guides fertilizer and foliar applications. Early tissue tests may also provide a basis for foliar feeding.

Samples from both affected and normal plants aid in diagnosing plant problems originating from nutrient deficiencies or excesses. Often the symptoms are present on leaves or other plant parts that are not analyzed in standard tests. Sampling affected parts and comparing the analysis with that of similar parts of normal plants often helps identify a nutritional problem. Sampling affected parts has been helpful in resolving some problems of phytotoxicity. It is very important to select tissues of the same age and position on the plant. If the affected parts are different than those used for standard sampling, the collection of additional samples following standard procedures can be helpful. While this makes for a large number of samples, the cost is usually justified by more certain problem resolution.

Precautions for Field Tissue Testing

In addition to the precautions outlined earlier, several others need to be observed for tissue testing, particularly for those tests run in the field.

Sampling Time of Day

Nitrate levels are affected by temperature and the time of day when samples are collected. Restricting collection to the period between 10:00 a.m. to 2:00 p.m. tends to give dependable results.

Storage of Samples

Because of nitrate breakdown, samples can be stored only for short periods. Whole petioles (with blades removed) can be stored in plastic bags on ice for about eight hours, but safe storage at room temperature is limited to 1.5-2.0 hours.

Nitrate Readings

Sap is not stored. It is expressed and nitrate readings are taken immediately.

Care of Kits

Chemicals or test strips for determining nitrates and other elements are subject to deterioration over time. Storing them in a cool, dry place or under conditions specified by the manufacturer can extend their usefulness. It is always worthwhile to run the tests with standards in order to calibrate the readings. Preparation of new standards each time increases accuracy.

Nitrate-N versus Nitrates

It is important to note whether kits are calibrated as nitrates or nitrate-N. If calibrated as nitrates, divide the values by 4.43 to obtain the equivalent nitrate-N value.

TABLE 5.1. Range of sufficiency values for total analysis of plant tissues

AGRONOMIC CROPS

Plant	Barley	Cassava	Corn	Corn	Corn	Cotton	Oat	Peanut	Peanut	Rice	Rice
Tissue	Tops	MRDL	Tops	LBL	EL	VEGS	Tops	Tops		MRDL	MRDL
Age-time	EB	VEG.	<12"	<Tas.	IS	FB	EB	BBL	EP	MTIL	PI
Number	25	25	15	12	12	25	25	50	25	25	25
						Analyses					
N %	1.75-3.0	5.0-6.0	3.5-5.0	3.0-3.5	2.7-4.0	3.0-4.3	2.0-3.0	3.5-4.5	3.5-4.5	2.0-3.6	2.6-3.2
P %	0.2-0.5	0.3-0.5	0.03-0.5	0.25-0.45	0.25-0.5	0.25-0.45	0.2-0.5	0.25-0.5	0.2-0.35	0.1-0.18	0.09-0.18
K %	1.5-3.0	1.2-2.0	2.5-4.0	2.0-2.5	1.7-3.0	0.9-2.0	1.5-3.0	1.7-3.0	1.7-3.0	1.2-2.4	1.0-2.2
Ca %	0.3-1.2	0.6-1.5	0.3-0.7	0.25-0.5	0.2-1.0	2.2-3.5	0.2-0.5	1.25-2.0	1.25-1.75	0.15-0.3	0.2-0.4
Mg %	0.15-0.5	0.25-0.5	0.15-0.45	0.13-0.3	0.2-1.0	0.3-0.9	0.15-0.5	0.3-0.8	0.3-0.8	0.15-0.3	0.2-0.3
S %	0.15-0.4	0.3-0.4	0.15-0.5	0.15-0.5	0.2-0.5		0.15-0.4	0.2-0.35	0.2-0.3		
B ppm	5-25	15-20	5-25	4-25	5-25	20-40	5-25	25-60	20-50	5-15	6-17
Cu ppm		7-15	5-20	3-15	6-20	5-25	5-25	5-20	10-50	8-25	8-25
Fe ppm		60-200	50-250	10-20	20-250	40-300	40-100	60-300	100-250	75-200	70-150
Mn ppm	25-100	50-250	20-300	15-300	20-200	30-300	25-100	60-350	100-350	200-800	150-800
Mo ppm	0.11-0.18		0.1-10	0.13-3.0	>0.2		0.2-0.3	0.1-5.0	0.1-5.0		
Zn ppm	15-70	40-100	20-60	15-60	25-100	20-100	15-70	25-60	20-50	25-50	18-50

116

Plant	Rye	Sorghum	Sorghum	Sorghum	Soybean	Sugar beet	Sugar cane	Tobacco	Wheat spring	Wheat winter
Tissue	Tops	Tops	MRDL	#3 leaf	MRDL	L	#3L	MRDL	Tops	#1 & #2L
Age-time		<12"	37-56d	BL	BPS	50-60d	3-5 mo	FL	EB	EB
Number	25	25	25	25	25	25	15	15	25	50
					Analyses					
N %	4.0-5.0	3.5-5.1	3.2-4.2	3.0-4.0	4.0-5.5	4.3-5.0	2.0-2.6	3.5-4.25	2.0-3.0	1.75-3.0
P %	0.5-0.65	0.3-0.6	0.13-0.25	0.15-0.25	0.25-0.5	0.45-1.1	0.1-0.3	0.27-0.5	0.2-0.5	0.2-0.5
K %	1.9-2.3	3.0-4.5	2.0-3.0	1.0-1.5	1.7-2.5	2.0-6.0	1.1-1.8	2.5-3.2	1.5-3.0	1.5-3.0
Ca %		0.9-1.3	0.15-0.9	0.2-0.5	0.35-2.0	0.5-1.5	0.2-0.5	1.5-3.5	0.2-0.5	0.2-0.5
Mg %	0.2-0.6	0.35-0.5	0.2-0.5	0.1-0.5	0.25-1.0	0.25-1.0	0.1-0.35	0.25-0.65	0.15-0.5	0.16-1.0
S %				0.2-0.4	0.2-0.4	>0.14	>0.14	0.25-0.5	0.15-0.4	
B ppm		4-13	1-10	1-6	21-55	31-200	4-30	20-50	6-10	
Cu ppm	<5	8-15	2-15	1-3	10-30		5-15	10-25	5-25	5-25
Fe ppm	14-45	160-250	55-200	40-80	50-350	60-140	40-250	50-200	25-100	10-300
Mn ppm	40-150	40-150	6-100	8-40	25-100	26-360	25-400	30-250	25-100	20-200
Mo ppm	0.2-2.0			1.0-5.0	1.0-5.0	0.20-2.0	0.05-4.0	0.4-0.6	0.09-0.18	
Zn ppm	30-60	30-60	20-40	7-16	21-50	10-80	20-100	20-80	15-70	21-70

TABLE 5.1. (continued)

FOLIAGE AND ORNAMENTALS

Plant	Aglaonema Chinese evergreen	Allamanda	Alstroemeria	Anthurium	Aphelandra Zebra pt.	Aralia, false	Asparagus fern	Asparagus Myers Sprengeri	Azalea	Azalea Indian hybrid	Baby's breath lflet	Begonia Rieger
Tissue Age	MRDL	MRDL	MRDL	LB+H-P MRDL	MRDL	MRDL	MRDCL	MRDCL	MRDL	MRDL	MRDL	MRDL
Number	12	25	20	15	25	25	15	15	50	50	50	20
						Analyses						
N %	3.0-3.8	2.0-4.0	3.7-5.6	1.6-3.0	2.0-3.0	2.5-3.5	1.5-3.1	1.5-2.5	1.5-2.5	2.0-3.0	4.3-6.0	3.5-6.0
P %	0.2-0.4	0.25-1.0	0.3-0.75	0.2-0.75	0.2-0.4	0.25-0.6	0.2-0.3	0.3-0.7	0.2-0.5	0.3-0.5	0.3-0.7	0.3-0.75
K %	3.0-4.5	2.0-4.0	3.7-4.0	1.0-3.5	1.1-2.0	1.0-3.5	2.2-3.5	2.0-3.0	0.5-1.5	1.5-2.5	3.5-4.5	2.5-6.0
Ca %	1.0-2.0	0.75-1.5	0.6-1.5	1.2-2.0	0.6-2.0	0.5-2.0	0.4-0.6	0.1-0.3	0.5-1.5	0.7-1.5	2.6-4.0	1.0-2.5
Mg %	0.3-0.6	0.25-1.0	0.2-0.5	0.5-1.0	0.5-1.0	0.25-0.4	0.1-0.3	0.1-0.3	0.25-1.0	0.25-0.5	0.4-1.0	0.3-0.7
S %	0.2-0.4	0.2-0.4	0.3-0.75	0.16-0.75	0.2-0.3	0.25-0.5	0.15-0.3	0.15-0.25	0.2-0.5	0.2-0.6	0.25-0.7	0.3-0.7
B ppm	25-75	25-75	13-50	25-75	35-50	25-50	20-40	20-40	25-50	25-50	25-75	20-70
Cu ppm	8-25	8-25	4-50	6-30	10-25	6-25	6-20	10-25	6-25	6-25	9-25	7-30
Fe ppm	50-250	50-200	150-300	50-300	50-300	50-300	50-300	50-300	50-250	50-200	50-200	50-250
Mn ppm	50-300	50-200	50-200	50-300	50-300	50-300	40-200	40-300	40-200	40-200	50-200	50-200
Zn ppm	20-75	20-75		20-50	20-75	20-75	25-75	25-75	20-75	20-75	25-75	25-75

Plant	Begonia wax leaf	Bird of paradise	Bougain-villea	Boxwood Japanese	Bromeliad Aechmea	Cactus Christmas	Caladium	Calathea	Carissa Natal plum	Carnation	Chalcas
Tissue	MRDL	B+M-P MRDL	MRDL	MRDL	MRDL NFL	MRDL	B+M UF	NRDL	MRDL	5th & 6th leaf pairs None flowering shoot	MRDL
Age											
Number	20	12	25	50	15	15	12	25	50	50	50
						-Analyses-					
N %	4.0-6.0	1.0-2.5	2.5-4.5	3.0-3.6	1.5-2.0	2.0-4.5	3.6-4.5	2.5-4.0	1.8-3.5	3.2-5.2	2.0-3.0
P %	0.3-0.75	0.2-0.4	0.25-0.75	0.3-0.5	0.3-0.7	0.6-1.0	0.3-0.7	0.2-0.5	0.18-0.6	0.25-0.8	0.25-0.5
K %	2.5-6.0	1.5-3.0	3.0-5.5	1.25-2.0	1.5-3.0	4.0-6.0	2.3-4.0	2.5-4.5	1.5-3.5	2.8-6.0	1.7-3.5
Ca %	1.0-2.5	0.1-3.0	1.0-2.0	1.0-2.0	0.5-1.0	0.8-1.5	1.0-1.5	0.5-1.5	1.0-3.0	1.0-2.0	0.8-1.5
Mg %	0.3-0.7	0.1-0.75	0.25-0.75	0.3-0.6	0.4-0.8	0.4-1.0	0.2-0.4	0.25-0.6	0.25-1.0	0.25-0.7	0.25-0.4
S %	0.3-0.7	0.2-0.4	0.2-0.5		0.16-0.25	0.25-0.5		0.2-0.4	0.2-0.4	0.25-0.8	0.25-0.4
B ppm	20-75	20-75	25-75		25-50	20-50	40-100	15-50	25-100	30-100	25-50
Cu ppm	7-30	8-30	8-25		7-30	10-30	5-10	6-25	6-25	8-30	7-25
Fe ppm	50-250	50-200	50-300		50-300	75-300	60-100	30-200	50-200	50-200	60-350
Mn ppm	50-200	45-200	50-200		50-300	60-300	50-210	30-200	50-250	50-200	50-250
Zn ppm	25-75	18-75	20-75		25-75	25-100	30-150	20-75	20-200	25-200	22-200

TABLE 5.1. Foliage and ornamentals (continued)

Plant	Chrysanthemum mum or pompon	Chrysanthemum mum or pompon	Coco plum	Coffee	Croton	Crown of thorns	Cyclamen	Dieffen-bachia	Dracaena godsef-fiana	Dracaena Janet Craig
Tissue	4th leaf from tip	4th leaf from tip	MRDL	MDDL	MRDL	MRDL	tops	B+MR-P MRDL	MRDL	MRDL
Age	To bud start	Bud start-harvest					50 days			
Number	25	25	50	25	12	25	15	10	15	15
					----Analyses----					
N %	4.0-6.0	3.5-5.0	2.0-3.0	2.5-3.5	1.5-3.0	2.0-4.0	2.0-2.7	2.7-3.3	1.8-3.0	2.0-3.0
P %	0.25-1.0	0.23-0.7	0.25-1.0	0.15-0.35	0.25-0.5	0.25-1.0	0.14-0.22	0.2-0.5	0.2-0.4	0.2-0.4
K %	4.0-6.0	3.5-5.0	1.0-2.5	2.0-3.0	1.3-3.0	1.5-4.0	2.2-5.7	3.5-4.5	1.2-2.5	2.5-4.0
Ca %	1.0-2.0	1.2-2.5	0.8-2.0	0.8-1.6	1.0-2.5	1.0-2.5		1.0-2.0	0.9-1.5	1.0-2.0
Mg %	0.25-1.0	0.25-1.0	0.25-1.0	0.3-0.5	0.3-1.0	0.25-1.0		0.3-0.75	0.2-0.5	0.3-0.6
S %	0.25-0.7	0.25-0.7	0.25-0.4	0.25-0.5	0.2-0.4	0.2-0.4		0.2-0.5	0.2-0.4	0.2-0.4
B ppm	25-75	25-75	25-75	25-75	25-75	25-100		15-50	20-50	16-50
Cu ppm	6-30	6-30	10-25	10-25	10-25	10-25		8-25	8-25	8-25
Fe ppm	50-250	50-250	50-200	90-300	50-200	50-200		50-300	40-300	50-300
Mn ppm	50-250	50-250	40-200	50-300	50-200	50-200		50-300	50-300	50-300
Mo ppm										
Zn ppm	20-75	20-75	20-75	15-75	20-75	20-75		20-75	20-75	20-75

Plant	Dracaena Corn plant	Dracaena Sanderana	Dracaena reflexa	Dracaena Warneckii	Eugenia	Fern Birdsnest	Fern Boston	Fern Leatherleaf	Fern Maidenhair	Fern Pteris	Ficus benjamina	Ficus decora
Tissue	MRDL	MRDL	MRDL	MRDL	MRDL	MRDL	MRDF	MRDF	MRDF	MRDF	MRDL	B+M-P MRDL
Age												
Number	15	15	15	15	50	50	15	15	25	15	50	10
N %	2.2-3.5	2.5-3.5	2.5-3.5	2.5-4.0	1.5-2.5	2.1-3.2	2.5-3.0	2.0-3.0	1.8-2.5	2.3-3.0	1.8-2.5	1.3-2.5
P %	0.15-0.4	0.2-0.4	0.22-0.4	0.2-0.5	0.4-0.8	0.3-0.5	0.25-0.7	0.25-0.5	0.3-0.6	0.21-0.3	0.1-0.5	0.1-0.5
K %	2.0-4.0	2.0-3.0	1.5-3.5	2.5-4.0	1.5-3.0	2.5-4.2	1.6-3.8	2.0-4.0	2.4-3.5	1.0-2.0	1.0-2.0	0.6-2.1
Ca %	1.0-2.5	1.5-2.5	1.5-2.5	0.9-1.2	1.0-2.5	0.5-1.0	0.8-2.5	0.5-1.0	0.12-0.3	2.0-3.0	1.0-3.0	0.3-1.2
Mg %	0.2-0.8	0.3-0.6	0.25-0.4	0.25-0.6	0.2-0.8	0.25-0.4	0.25-1.0	0.2-0.8	0.25-0.4	0.25-0.4	0.4-1.0	0.2-0.5
S %	0.2-0.7	0.2-0.4		0.2-0.4	0.2-0.4	0.2-0.35	0.2-0.5	0.2-0.5	0.2-0.4		0.15-0.5	0.15-0.5
B ppm	20-50	20-50	20-35	18-50	25-75	15-50	20-70	20-50	25-50	20-30	30-75	20-50
Cu ppm	8-25	10-25	6-25	8-25	10-25	8-20	6-25	5-25	10-25	6-30	8-25	8-25
Fe ppm	50-300	50-300	50-200	50-300	50-200	50-300	50-300	20-200	50-300	40-300	30-200	30-200
Mn ppm	50-300	50-300	40-200	50-300	50-200	30-300	40-200	30-300	30-300	70-300	25-200	20-200
Mo ppm												
Zn ppm	20-75	20-75	20-75	20-75	20-75	20-75	20-75	20-75	25-75	25-75	15-75	15-75

-Analyses-

TABLE 5.1. Foliage and ornamentals (continued)

Plant	Ficus Fiddle-leaf	Gardenia	Geranium	Gerbera Transvaal daisy	Gladiolus	Gloxinia	Hibiscus	Holly American	Holly Chinese	Holly Japanese	Hydrangea	Ixora
Tissue	B+M-p	MRDL	MRDL	MRDL	MRDL	MRDL	MRDL	MRDL	MRDL	MRDL	MRDL	MRDL
Age Number	10	25	15	15	12	15	15	50	50	50	15	25
							----Analyses----					
N %	1.3-2.3	1.5-3.0	3.5-4.8	1.5-3.5	3.0-5.5	3.0-5.0	2.5-4.5	1.5-2.2	1.7-2.4	1.8-2.8	3.0-5.5	1.8-2.3
P %	0.1-0.5	0.16-0.4	0.4-0.7	0.2-0.5	0.25-1.0	0.25-0.7	0.25-1.0	0.1-0.2	0.12-0.3	0.15-1.0	0.25-0.7	0.15-1.0
K %	0.6-2.1	1.0-3.0	2.5-4.3	2.5-4.5	2.5-4.0	2.5-5.0	1.5-3.0	1.4-2.8	0.8-2.2	1.0-2.0	2.2-5.0	1.0-2.5
Ca %	0.3-1.2	0.5-1.3	0.8-1.2	1.0-3.5	0.5-4.5	1.0-3.0	1.0-3.0	0.5-1.0	0.7-1.5	0.25-0.8	0.6-1.8	0.8-2.0
Mg %	0.2-0.5	0.25-1.0	0.2-0.5	0.2-0.7	0.15-0.3	0.35-0.7	0.25-0.8	0.2-0.6	0.3-1.0	0.25-0.8	0.22-0.5	0.2-1.0
S %	0.15-0.5	0.2-0.4	0.25-0.7	0.25-0.7		0.25-0.7	0.2-0.5			0.2-0.4	0.2-0.7	0.2-0.4
B ppm	20-50	25-75	30-200	20-60	25-100	25-50	25-100	35-65	30-50	25-100	20-50	25-100
Cu ppm	8-25	6-25	7-25	6-25	8-20	8-25	6-25	10-25	15-25	10-25	6-25	10-25
Fe ppm	30-200	60-250	100-250	50-200	50-200	50-200	50-200	100-250	100-200	65-300	50-300	65-250
Mn ppm	20-200	50-250	40-200	40-250	50-200	50-300	40-200	50-125	75-150	50-200	50-300	50-200
Mo ppm				0.2-0.6								
Zn ppm	15-75	20-75	18-200	25-200	20-75	20-50	20-75			20-75	20-75	20-75

Plant	Jasmine	Juniper	Kalanchoe	Leea Hawaiian holly	Ligustrum	Lily, day	Lily, Easter	Lipstick plant	Liriope	Malpighia	Mande-villa	Maranta Rabbits foot
Tissue	max	Ti-S RM6S	#4 leaf from tip	Lfit	MRDL	MRDL	MRDL	MRDL	MRDL	MRDCL	MRDL	MRDL
Age	MRDL			MRDL								
Number	50	25	15	25	25	15	15	25	25	25	25	25
N %	2.0-4.0	1.5-2.5	2.2-4.5	2.2-3.3	2.2-3.0	3.0-5.0	3.3-4.8	2.1-2.8	2.0-3.0	2.0-3.5	1.9-3.0	2.0-3.0
P %	0.18-0.5	0.2-0.75	0.25-1.0	0.19-0.5	0.2-0.5	0.25-0.5	0.25-0.7	0.2-0.4	0.25-0.35	0.15-0.5	0.20-0.5	0.25-0.5
K %	1.3-2.5	1.0-2.5	3.0-4.0	1.5-2.8	1.6-3.5	2.5-4.0	3.3-5.0	2.5-3.3	2.0-2.9	1.5-3.0	2.0-4.0	3.0-5.5
Ca %	0.7-1.5	0.8-1.5	2.0-4.0	1.2-2.0	0.7-1.5	0.5-1.0	0.6-1.5	0.8-1.6	0.9-1.5	1.0-3.5	0.8-1.5	0.6-1.5
Mg %	0.25-1.0	0.25-0.8	0.6-1.5	0.25-0.8	0.15-0.3	0.2-0.5	0.2-0.7	0.25-0.4	0.15-0.25	0.25-0.8	0.25-0.5	0.25-1.0
S %	0.2-0.4	0.2-0.45		0.2-0.5	0.2-0.4	0.25-0.7	0.25-0.7	0.2-0.3		0.2-0.4	0.2-0.4	0.2-0.5
B ppm	25-75	20-60	6-10	15-50	20-60	25-75	25-75	25-50	20-35	25-75	25-75	25-50
Cu ppm	10-25	8-25	4-10	10-25	5-25	8-25	8-25	10-25	6-15	6-25	8-25	8-25
Fe ppm	50-200	40-200	55-100	30-300	50-200	50-300	60-200	50-200	50-200	50-200	50-200	60-300
Mn ppm	40-200	25-200	70-100	30-200	30-250	25-300	35-200	50-300	25-75	25-200	25-200	50-200
Mo ppm												
Zn ppm	20-75	20-75	30-100	20-75	20-75	20-75	20-75	25-75	19-35	20-75	20-75	20-75

Analyses

TABLE 5.1. Foliage and ornamentals (continued)

Plant	Marble queen Pothos	Monkey puzzle tree	Nephthytis	Norfolk island pine	Olive, Black	Orchid Cattleya Cymbidium	Orchid Lady-slipper	Orchid Moth or Phalaenopsis	Palm Areca	Palm Bamboo Seifritzii	Palm Parlor
Tissue	MRDL	MRDL	MRDL	MRDL	MRDL	MRDL	MRDL	MRDL	MLF-LP MRDL	MLF-LP MRDL	MLF-LP MRDL
Age Number	25	25	25	25	25	12	12	12	15	15	15
						──Analyses──					
N %	2.7-3.5	1.2-2.5	2.5-3.5	1.5-2.8	1.6-3.0	1.5-2.5	2.3-3.5	2.0-3.5	2.5-3.5	2.5-3.5	2.5-3.5
P %	0.2-0.5	0.16-0.3	0.2-0.5	0.2-0.3	0.15-0.75	0.13-0.75	0.02-0.7	0.2-0.7	0.15-0.8	0.15-0.3	0.15-0.3
K %	3.0-4.5	1.5-2.5	3.0-4.5	1.5-2.5	0.7-3.5	2.0-3.5	2.0-3.5	4.0-6.0	1.4-4.0	1.6-2.8	1.6-3.0
Ca %	1.0-2.0	0.7-1.5	0.4-1.5	0.7-1.5	0.25-1.0	0.5-2.0	0.75-2.0	1.5-2.5	1.0-2.5	1.0-2.5	0.8-2.5
Mg %	0.3-1.0	0.2-0.5	0.3-0.6	0.2-0.5	0.25-1.0	0.3-0.7	0.2-0.7	0.4-1.0	0.25-0.8	0.25-0.8	0.25-0.8
S %		0.13-0.25	0.2-0.5	0.15-0.25	0.2-0.75	0.15-0.75	0.2-0.7	0.2-0.7	0.2-0.75	0.2-0.4	0.2-0.4
B ppm	20-60	15-40	25-50	15-40	25-75	25-75	25-75	25-75	15-60	25-60	25-60
Cu ppm	6-25	6-25	10-25	6-25	5-25	5-20	5-20	5-20	6-25	6-25	6-25
Fe ppm	50-300	50-300	50-300	50-300	50-200	50-200	50-200	75-200	50-250	50-300	50-300
Mn ppm	50-300	30-250	50-300	30-250	40-200	40-200	50-200	100-200	50-250	50-250	50-250
Mo ppm											
Zn ppm	20-75	20-75		25-75	20-75	25-75	25-75	20-75	25-75	25-75	25-75

Plant	Palm Kentia	Palm Pony tail	Palm Lady or Rhapis	Palm Roebel-lini	Peperomia	Philo-dendron cordatum	Philo-dendron hastatum	Philodendron panduriforme	Philodendron pertusum Monstera	Philo-dendron Selloum
Tissue	MLF-LP	MLF	MLF-LP	MLF-LP	MRDL	MRDL	B+M-P	B+M-P	B+M-P	B+M-P
Age	MRDL	MRDL	MRDL	MRDL				MRDL	MRDL	MRDL
Number	15	12	15	15	15	12	8	8	8	8
							--Analyses--			
N %	2.2-2.8	1.5-2.0	1.8-2.8	2.0-2.8	2.9-4.5	2.5-3.5	2.5-4.0	2.5-4.5	3.0-5.0	3.5-5.0
P %	0.12-0.25	0.14-0.2	0.15-0.8	0.16-0.4	0.25-1.0	0.2-0.5	0.25-0.6	0.23-0.45	0.2-0.4	0.25-0.5
K %	1.3-2.5	1.7-2.8	1.5-2.5	1.2-2.5	4.0-6.0	3.0-6.0	3.0-4.5	2.0-3.7	2.5-4.5	2.0-4.0
Ca %	1.0-2.0	1.0-2.0	0.41-1.0	0.6-1.5	1.0-4.0	0.35-2.5	1.0-2.5	1.0-2.0	1.5-2.5	1.0-2.5
Mg %	0.22-0.3	0.2-0.3	0.2-0.3	0.18-0.3	0.4-1.2	0.25-0.5	0.3-0.5	0.25-0.5	0.25-0.5	0.25-1.0
S %			0.15-0.75		0.25-0.75	0.21-0.4	0.25-0.6	0.2-0.5	0.2-0.5	
B ppm	19-30	20-35	16-75	16-30	25-50	25-75	30-70	20-50	17-60	10-75
Cu ppm	6-10	8-25	7-25	6-20	7-25	8-25	10-25	7-25	7-25	6-25
Fe ppm	100-300	60-200	80-300	50-200	50-300	40-200	50-200	60-200	50-200	50-300
Mn ppm	60-150	25-200	50-250	25-200	50-300	50-200	50-200	40-200	50-200	50-300
Mo ppm										
Zn ppm	20-75	25-75	20-75	30-100	25-75	25-75	18-25	25-50	25-75	20-75

TABLE 5.1. Foliage and ornamentals (continued)

Plant	Pitto-sporum	Podo-carpus	Poin-settia	Rhodo-dendron	Rose	Salvia	Sanse-vieria	Schef-flera Umbrella	Snap-dragon	Spathi-phyllum	Spathi-phyllum	Spider plant
Tissue	MRDL	MRDL	MRDL	L Dormant	C FIVE LFT MRDL	MRDL	MRDL	L+MR-P MRDL	MRDL	B+MR-P MRDLP<4MO	B+MR-P MRDLP>4MO	MRDL
Age Number	25	25	15	15	10	25	15	10	25	12	8	15
						-Analyses-						
N %	1.3-3.0	2.0-3.5	4.0-6.0	1.6-2.2	3.0-5.0	3.0-4.5	1.7-3.0	2.5-3.5	3.8-5.0	3.8-5.0	3.3-4.5	1.7-3.0
P %	0.25-1.0	0.25-1.0	0.3-0.5	0.2-0.3	0.25-0.5	0.3-0.7	0.15-0.4	0.2-0.5	0.3-0.5	0.25-1.0	0.2-0.8	0.15-0.4
K %	1.4-3.5	0.8-2.0	1.5-3.5	0.5-1.0	1.5-3.0	3.5-5.0	2.0-3.0	2.3-4.0	2.0-3.0	4.0-6.0	2.3-4.0	2.5-5.0
Ca %	0.8-2.5	1.0-2.0	0.7-2.0	0.8-1.4	1.0-2.0	1.5-2.5	1.0-2.0	1.0-1.5	1.0-1.5	1.0-2.0	1.0-2.0	1.0-2.5
Mg %	0.18-1.0	0.25-0.8	0.3-1.0	0.2-0.3	0.25-0.5	0.25-0.6	0.3-0.6	0.25-0.75		0.25-0.5	0.25-0.5	0.25-1.5
S %	0.2-0.4	0.2-0.4	0.25-0.7		0.25-0.7		0.2-0.5	0.2-0.8		0.25-0.5	0.2-0.5	
B ppm	20-75	20-75	30-100	30-50	30-60	25-75	20-50	20-60		25-70	25-70	25-40
Cu ppm	6-25	10-25	3-25	1-5	7-25	7-25	10-25	10-25		8-25	8-25	8-25
Fe ppm	40-200	30-200	100-300	100-250	60-200	60-300	50-300	50-300		50-300	50-300	60-150
Mn ppm	25-200	25-200	45-300	500-1200	30-200	30-200	40-300	40-300		40-300	40-300	50-75
Mo ppm			1.0-5.0		0.1-0.9	2-4						
Zn ppm	20-75	20-75	25-100	20-40	18-75	25-75	25-75	20-75		25-75	25-75	25-75

Plant	Statice	Syringa	Taxus	Viburnum	Violet African	Yucca
Tissue	MRDCL	CSL	CSL	MRDL	MRDL	MRDL
Age Number	20	15	75	20	15	12
N %	3.5-6.0	1.6-2.5	2.0-4.0	1.5-2.5	3.0-6.0	1.8-2.5
P %	0.3-0.7	0.25-0.4	0.3-0.5	0.15-0.4	0.3-0.7	0.15-0.8
K %	3.0-5.0	1.0-1.8	1.0-2.0	0.9-2.0	3.0-6.5	1.2-2.8
Ca %	0.5-1.0	0.6-1.2	0.6-1.0	0.6-1.5	1.0-2.0	1.0-2.5
Mg %	0.5-1.2	0.2-0.4	0.2-0.3	0.25-1.0	0.35-0.75	0.2-0.6
S %				0.2-0.4	0.3-0.7	0.2-0.8
B ppm	20-40	18-40	20-35	20-75	25-75	18-60
Cu ppm	7-25	8-25	10-20	7-25	8-35	8-25
Fe ppm	50-200	75-300	75-250	30-200	50-200	25-200
Mn ppm	50-200	30-300	100-500	30-200	40-200	40-200
Mo ppm		1-4	1-3	1-3		
Zn ppm	25-75	25-75	25-100	20-75	25-75	20-75

Analyses

127

TABLE 5.1. (continued)

FORAGE, GRASS, HAY, AND TURF

Plant	Alfalfa	Birdsfoot trefoil	Clover Alsike	Clover White or Ladino	Clover Red	Clover Subterranean	Grains small	Grass Bermuda Fairways	Grass Bermuda Greens	Grass Broome	Grass Coastal Bermuda	Grass Creeping Bent
Tissue	Top 6 in.	L	Top	L	Top	MRDL	MRDL	Clippings	Clippings	FFDS+L	Tops	Clippings
Age	BBL	FFL	FFL	BBL	BBL	BBL	BHead			MIDCUTS	4-5 week	6\1-8\30
Number	12	50	20	50	15	75	50			25	15	
						‑‑‑‑Analyses‑‑‑‑						
N %	4.5-5.0	4.0-4.5		4.5-5.0	3.0-4.5	3.0-3.5	2.0-3.0	3.0-5.0	4.0-6.0	2.0-3.5	2.2-4.0	4.5-6.0
P %	0.26-0.7	0.28-0.36	0.25-0.5	0.36-0.45	0.28-0.6	0.25-0.3	0.2-0.4	0.15-0.5	0.25-0.6	0.25-0.35	0.25-0.6	0.3-0.6
K %	2.0-3.5	1.6-2.6	1.5-3.0	2.0-2.5	1.8-3.0	1.0-1.5	1.5-3.0	1.0-4.0	1.5-4.0	2.0-3.5	1.8-3.0	2.2-2.6
Ca %	1.8-3.0	1.7-2.0	1.0-1.8	0.5-1.0	2.0-2.6	1.0-1.5	0.2-0.5	0.5-1.0	0.5-1.0	0.25-0.4	0.25-0.5	0.5-0.75
Mg %	0.3-1.0	0.4-0.6	0.3-0.6	0.2-0.3	0.21-0.6	0.2-0.5	0.15-0.5	0.13-0.5	0.13-0.4	0.14-0.3	0.13-0.3	0.25-0.75
S %	0.26-0.5			0.25-0.5	0.26-0.3	0.2-0.5		0.15-0.5	0.2-0.5	0.17-0.3	0.18-0.5	0.25-0.3
B ppm	30-80	30-75	15-50	25-50	30-80	>25	3-20	6-30	5-30	10-20	6-30	8-20
Cu ppm	7-30	6-10	3-15	5-8	8-15	7-13	5-25	5-50	5-50	5-10	5-25	8-30
Fe ppm	30-250	50-80	50-100	50-100	30-250	50-200	25-100	50-350	50-350	50-100	50-350	100-300
Mn ppm	31-100	50-80	40-100	25-100	30-120	50-120	25-100	25-300	25-300	40-80	25-300	50-100
Zn ppm	1.0-5.0			0.15-0.25	0.5-1.0	2						
Zn ppm	21-70	30-50	15-80	15-25	18-80	25-50	20-70	20-75	20-75	20-50	20-50	25-75

Plant	Grass Kentucky Blue	Grass Orchard Cocksfoot	Grass Pangola	Grass Perennial Rye	Grass Sudan Sorg-sudan	Grass St. Augustine	Grass Tall Fescue	Millet	Stylo	Timothy
Tissue	Clippings	Tops	Tops	Tops	Tops	Clippings	Tops	Tops	Tops	Tops
Age-time	4-6 wks	3-4 wks	4-5 wks	Vegetative	4-5 wks		5-6 wks	4-5 wks		EARANT
Number		25	25	25	15		20	20	20	25
Analysis										
N %	2.6-3.5	3.2-4.2	1.7-2.5	4.5-5.0	2.0-3.5	1.9-3.0	3.4-3.8	2.5-4.0	3.0-3.5	0.6-1.7
P %	0.28-0.4	0.23-0.35	0.16-0.26	0.35-0.4	0.2-0.35	0.2-0.5	0.34-0.45	0.22-0.4	0.2-0.3	0.15-0.3
K %	2.0-3.0	2.6-3.5	1.6-2.2	2.0-2.5	1.9-3.5	2.5-4.0	3.0-4.0	2.3-4.5	0.6-1.2	1.2-1.8
Ca %	0.5-0.9		0.2-0.3	0.25-0.3		0.3-0.5			1.6-2.0	0.1-0.35
Mg %	0.15-0.3			0.16-0.2		0.15-0.25			0.3-0.5	0.1-0.3
S %	0.2-0.25			0.27-0.32				0.14-0.25	0.2-0.3	
B ppm	8-12			9-17		5-10				2-10
Cu ppm	3-5			6-7		20-20				7-25
Fe ppm	50-200					50-300				25-60
Mn ppm	50-150			40-60		40-200		30-43		12-35
Mo ppm	0.5-1.5			2-10						25-75
Zn ppm	20-50			14-20		20-75			35-50	

TABLE 5.1. (continued)

FRUITS AND NUTS

Plant	Almond	Apple	Apricot shipping	Apricot canning	Avocado	Banana	Banana	Blueberry Highbush	Blueberry Rabbiteye	Cashew #4 Leaf	Cherry Sour	Cherry Sweet
Tissue	FEL	FEL	FEL	FEL	FEL	6"MIDL	6"MIDL	FEL MSCS	FEL MSCS	FEL MB	FEL MSCS	FEL MSCS
Age	RMGS	MS CS	MS CS	MS CS	NFT	6-9mos.	Harvest	Jul-Aug	Jul-Aug		Jul-Aug	Jul-Aug
Number	50	50	50	50	50	12	12	75	75		50	50
					----Analyses----							
N %	2.2-2.5	1.9-2.0	2.0-2.5	2.5-3.0	1.6-2.0	3.5-4.5	2.5-3.0	1.5-2.2	1.2-1.7	1.65-2.75	2.6-3.0	2.1-3.0
P %	0.1-0.3	0.14-0.4	0.13-0.56	0.13-0.35	0.08-0.25	0.2-0.4	0.18-0.4	0.2-0.5	0.08-0.2	0.16-0.25	0.16-0.22	0.16-0.5
K %	1.4-	1.5-2.0	2.5-3.0	2.5-3.0	0.75-2.0	3.8-5.0	2.3-4.0	0.5-0.9	0.35-0.6	0.9-1.45	1.6-2.1	2.5-3.0
Ca %	2.0-	1.2-1.6	1.6-2.5	1.6-2.5	1.0-3.0	0.8-1.5	0.7-1.4	0.5-0.8	0.25-0.7	0.03-0.12	1.5-2.6	2.0-3.0
Mg %	0.25-	0.25-0.4	0.3-1.2	0.3-1.2	0.25-0.8	0.25-0.8	0.25-0.4	0.25-0.4	0.14-0.2	0.02-0.05	0.3-0.75	0.3-0.8
S %		0.2-0.4			0.2-0.6	0.25-0.8	0.26-0.5	0.2-0.4	0.11-0.25			
B ppm	30-60	25-50	25-70	20-70	50-100	10-50	15-50	25-75	12-35		20-55	20-100
Cu ppm	4-	6-50	5-25	5-25	5-15	6-25	6-30	4-15	4-10		8-28	5-50
Fe ppm		50-300	70-150	70-150	50-200	75-300	100-300	36-200	27-70		100-200	100-200
Mn ppm	20-	25-300	25-100	25-100	30-500	100-1000	200-2000	40-250	25-100		40-60	90-200
Mo ppm		<0.1			0.05-1.0							
Zn ppm	18-	20-100	20-100	20-60	30-150	20-200	13-50	20-100	10-25		20-50	20-50

Plant	Cranberry	Currant Black	Fig	Grape Vitis labrusca	Grape Vitis vinifera	Grape Vitis vinifera	Grape Muscadine	Grapefruit	Grapefruit	Hazelnut	Lemon	Lime Persian
Tissue	FEL CS	FEL MSCS	FEL MSCS	PT OBFC FBL	PT OBFC FBL	FEL OBC	FEL OBC	NFL	LBF	FEL MSCS	NFL	NFL VEGSH
Age-Time	Jul-Aug		Jul-Aug			June-July	M-LS	5-7 mos	5-7 mos	July	5-7 mos	
Number	50	50	25	50	50	15	15	25	25	25	25	30
						---Analyses---						
N %	0.8-1.2	2.0-2.5	2.0-2.5	1.6-2.8	1.7-3.0	2.0-2.3	1.65-2.15	2.0-2.2	2.0-2.6	2.3-2.6	2.0-2.6	2.4-3.0
P %	0.1-0.25	0.1-0.3	0.1-0.3	0.3-0.6	0.15-0.5	0.3-0.4	0.12-0.18	0.12-0.16	0.13-0.5	0.16-0.4	0.12-0.16	0.15-0.5
K %	0.4-0.8	1.4-1.7	1.0-	2.6-5.0	1.5-2.0	1.3-1.4	0.8-1.2	0.7-1.09	0.8-2.2	0.7-2.4	0.7-1.09	1.6-2.5
Ca %	0.5-0.9	1.3-2.5	3.0-	0.42-1.3	1.0-3.0	2.0-2.5	0.7-1.1	3.0-5.5	1.5-5.5	1.0-2.5	3.0-5.5	1.5-5.0
Mg %	0.15-0.3	0.2-0.5	0.75-	0.13-0.4	0.3-1.5	0.25-0.5	0.15-0.25	0.26-0.6	0.3-0.6	0.25-0.5	0.26-0.6	0.25-1.0
S %	0.1-0.15							0.20-0.3	0.15-0.5		0.20-0.3	0.15-0.5
B ppm	15-50	4-		25-50	30-100	25-70	15-25	31-100	31-100	31-75	31-100	30-100
Cu ppm	5-15		4-			5-50	5-10	5-16	5-15	4-50	5-16	5-100
Fe ppm	60-125	60-300		50-100	40-300	60-175	60-120	60-100	60-200	50-350	60-100	60-200
Mn ppm	75-200	50-100	20-	18-100	30-150	30-300	60-120	25-200	25-200	25-500	25-200	20-200
Mo ppm						0.15-0.35	0.15-0.35	0.1-0.29			0.1-0.29	
Zn ppm	25-40			20-30	25-100	25-100	18-35	25-100	25-200	15-80	25-100	20-100

TABLE 5.1. Fruits and nuts (continued)

	Macadamia	Mandarin Tangerine	Mango	Olive	Orange Navel &	Orange valencia	Papaya	Peach	Pear	Pecan	Pineapple	Plum or Prune
Plant												
Tissue	MSPF	VEGSH	MRFDL	FEL M	LBF	NFTL	P	MS L	FEL MSCS	LPRMPTG	DL-WHB	MSH
Age-Time		MRFD	POBL		5-7 mos	5-7 mos	MRFD	Mid-sum	Mid-sum	56-84days	INFST	Mid-sum
Number	50	30	15	50	30	30	15	25	50	25	20	25
						Analyses						
N %	1.5-2.5	3.0-3.4	1.0-1.5	1.5-2.5	2.2-3.5	2.4-2.6	1.01-2.5	3.0-3.5	2.2-2.8	2.7-3.5	1.5-1.7	2.4-3.0
P %	0.1-0.3	0.15-0.25	0.08-0.25	0.1-0.3	0.12-0.5	0.12-0.16	0.22-0.4	0.14-0.25	0.11-0.25	0.14-0.3	-0.1	0.14-0.25
K %	0.5-1.5	0.9-1.1	0.4-0.9	0.9-1.2	1.2-3.0	0.7-1.09	3.3-5.5	2.0-3.0	1.0-2.0	1.25-2.5	2.2-3.0	1.6-3.0
Ca %	0.5-1.0	0.17-0.44	2.0-5.0	1.0-	1.1-4.0	3.0-5.5	1.0-3.0	1.8-2.7	1.0-1.5	1.0-1.75	0.8-1.2	1.5-3.0
Mg %	0.1-0.3		0.2-0.5	0.2-	0.3-0.5	0.26-0.6	0.4-1.2	0.3-0.8	0.25-0.5	0.3-0.6	-0.3	0.3-0.8
S %	0.1-0.25					0.28-0.3			0.2-0.4	-0.2		
B ppm	25-50	31-100	25-150	20-75	25-100	31-100	20-30	20-60	20-70	15-50	30-	25-60
Cu ppm	6-12		7-25		6-100	5-16	4-10	5-16	5-20	6-30	-10	6-16
Fe ppm	25-200		50-250	25-	60-150	60-100	25-100	100-250	60-250	50-300	100-200	100-250
Mn ppm	100-400		50-200		25-200	25-100	20-150	40-160	30-100	200-500	50-200	40-160
Mo ppm	0.5-2.5					0.1-0.29						20-200
Zn ppm	15-50	5-29	20-75	25-	25-200	25-100	15-40	20-50	25-200	50-100	-20	20-50

Plant	Raspberry	Strawberry	Walnut
Tissue	MRFOL	MRFOL	MRFM CLF
Age-Time	FLBST	FL	Jul-Aug
Number	50	25	25
	---------------------Analyses---------------------		
N %	2.5-4.0	2.5-4.0	2.5-3.25
P %	0.3-0.5	0.25-1.0	0.12-0.3
K %	1.5-3.0	1.3-3.0	1.2-3.0
Ca %	0.8-1.5	1.0-2.5	-1.0
Mg %	-0.3	0.25-1.0	0.3-1.0
B ppm	25-75	25-50	35-100
Cu ppm	3-50	6-25	4-20
Fe ppm	50-250	50-200	
Mn ppm	50-250	50-200	30-300
Zn ppm	25-100	20-75	22-25

TABLE 5.1. (continued)

LANDSCAPE AND FOREST TREES

Plant	Fir, Douglas	Larch, Japanese	Maple, Sugar	Pine, Slash	Spruce, Red
Tissue	N TMLB	N US MT	MID1\3CRL	FFMNCY	N BLC
			MT Trees		So. Side
Age-Time	2-5 yrs	Late-sum	Aug 15-Sept	<2YR	Mid-Oct.
	Oct-Nov.				
Number	50	50	20	50	50
----Analyses----					
N %	1.5-2.3	2.4-3.0	2.5-3.0	1.5-4.0	0.85-1.35
P %	0.18-0.35	0.2-0.3	0.13-0.22	0.15-0.8	0.13-0.27
K %	0.75-1.1	1.1-1.3	0.6-1.0	1.0-3.0	0.55-1.15
Ca %	0.3-0.5	0.22-0.25	0.4-1.3	0.25-0.9	0.21-0.65
Mg %	0.09-0.15		0.1-0.25	0.12-0.5	0.07-0.12
S %	0.15-0.25			0.15-	
B ppm	4-15			21-50	15-42
Cu ppm	3-12		5-15	6-30	4-16
Fe ppm	70-200			30-200	25-60
Mn ppm	200-600		250-2000	50-400	50-260
Mo ppm	0.02-0.25		0.05-0.2		1.7-2.7
Zn ppm	25-45		20-35	20-100	25-55
Cl ppm	100-1000				
Na ppm			100-700		

VEGETABLES

Plant	Asparagus	Asparagus	Bean, Snap	Beet, Table	Broccoli	Brussels Sprouts	Cabbage	Cabbage	Cabbage	Cabbage	Cabbage	Cabbage, Chinese	Cabbage, Chinese
Tissue	Top 20"FR	Top 6"FR	MRFDUTRL	MRFDL	MRFDL	MRFDL	Tops	WRL	WRL	WRL M	FFDL	FFDL
Age-Time	Aug-Sept	FFDF			Heading	Mature	2-6wks	2-3mos	Mature	Mature	8L stage	Mature
Number	10	15	10	20	12	12	15	12	12	15	12	12
						Analyses							
N %	2.5-4.0	2.5-4.5	5.0-6.0	4.0-5.5	3.2-5.5	3.1-5.5	3.0-5.0	3.6-5.0	3.5-4.8	2.0-4.5	4.5-5.5	3.5-4.0
P %	0.25-0.5	0.2-0.35	0.35-0.75	0.25-0.5	0.3-0.75	0.3-0.75	0.35-0.75	0.33-0.75	0.3-0.65	0.25-1.0	0.5-0.6	0.4-0.6
K %	1.5-2.8	1.7-3.0	2.25-4.0	3.0-4.5	2.0-4.0	2.0-4.0	3.5-6.0	3.0-5.0	2.0-4.0	3.0-5.5	7.5-9.0	4.5-7.5
Ca %	0.6-1.0	0.4-1.0	1.5-2.5	2.5-3.5	1.0-2.5	1.0-2.5	3.0-4.5	1.1-3.0	1.3-3.5	1.0-2.0	3.0-5.5	2.0-6.0
Mg %	0.25-0.3	0.15-0.3	0.3-1.0	0.3-1.0	0.25-0.75	0.25-0.75	0.5-2.0	0.4-0.75	0.25-0.8	0.25-1.0	0.35-0.5	0.3-0.7
S %	0.25-0.5	0.25-0.5			0.3-0.75	0.3-0.75		0.3-0.75	0.3-0.75	0.25-1.0		
B ppm	40-100	30-50	20-75	30-85	30-100	30-100	25-75	25-75	30-100	25-75	25-75	30-100
Cu ppm	5-25	10-50	7-30	5-15	5-15	5-15	5-15	5-15	5-15	6-15	5-25	5-25
Fe ppm	40-250	50-300	50-300	50-200	70-300	60-300	30-200	30-200	30-200	40-200	30-200	40-200
Mn ppm	25-200	50-200	50-300	50-250	25-200	25-200	50-200	25-100	25-200	40-200	25-200	25-200
Mo ppm								0.4-0.7	0.4-1.0			
Zn ppm	20-100	15-30	20-75	20-75	35-100	25-100	25-75	20-75	20-75	20-75	30-75	20-75

135

TABLE 5.1. Vegetables (continued)

Plant	...Cantaloupe...	...Muskmelon...	Carrot	Carrot	Cauli-flower	Celery	Celery	Collards	Corn, Sweet	Corn, Sweet	Corn, Sweet	Corn, Sweet
Tissue	FFDL	FFDL	FFDL	OLDL	MRFDL	Petioles -MRFDL	Petioles OL-1'B&TN	FFDL	5thLFT-UN	5thLFT-UN	5thLFT-UN	5thLFT-UN
Age-Time	FLST-SMFR	SMFR-HAR	Mid-grown	Mature	Heading	6wks	Mature	FFDL	12-20"	5-6wks	Tassel start	Full T Silk ST
Number	12	12	15	15	12	12	12	12	8	8	8	8
					Analyses							
N %	4.5-5.5	4.0-5.0	2.1-3.5	3.0-3.5	3.3-4.5	1.6-2.0	0.7-1.5	4.0-5.0	4.0-4.5	3.5-4.5	2.7-3.5	2.5-3.0
P %	0.3-0.8	0.25-0.6	0.2-0.5	0.2-0.4	0.33-0.8	0.3-0.6	0.25-0.5	0.3-0.7	0.6-1.0	0.5-0.8	0.4-0.7	0.3-0.6
K %	4.0-5.0	3.5-4.5	2.8-4.0	2.9-3.5	2.6-4.2	8.6-10.0	7.0-9.5	3.0-4.5	3.5-4.5	2.9-3.8	2.5-3.5	1.5-2.5
Ca %	2.3-3.0	2.5-3.2	1.4-3.0	1.0-2.0	2.0-3.5	2.2-3.5	2.2-3.5	3.0-4.0	0.5-0.8	0.5-0.9	0.7-1.0	0.6-1.1
Mg %	0.35-0.8	0.35-0.8	0.3-0.5	0.25-0.6	0.27-0.5	0.25-0.5	0.3-0.6	0.25-0.75	0.2-0.5	0.2-0.5	0.2-0.5	0.2-0.5
S %	0.25-1.4	0.23-1.2							0.2-0.75	0.2-0.75	0.2-0.75	0.2-0.75
B ppm	25-60	25-60	30-100	30-75	30-100	25-50	25-60	30-100	8-25	8-25	8-25	8-25
Cu ppm	7-30	7-30	5-15	5-15	4-15	5-15	5-15	4-20	5-25	5-25	5-25	5-25
Fe ppm	50-300	50-300	50-300	50-300	30-200	30-100	22-100	50-150	50-300	50-300	50-300	50-300
Mn ppm	50-250	50-200	60-200	50-200	25-250	10-100	10-100	30-250	31-300	0.31-300	31-300	31-300
Mo ppm			0.5-1.5	0.5-1.4	0.5-0.8				0.9-10	0.9-10	0.9-10	0.9-10
Zn ppm	20-200	20-200	25-250	20-250	20-250	25-100	10-100	20-100	20-150	20-150	20-150	20-150

136

Plant	Corn, Sweet	Cucumber, Field	Cucumber, Field	Cucumber, European	Eggplant	Endive or Escarole	Garlic	Garlic	Garlic	Horse-radish	Kale	Kohlrabi
Tissue	5thL-UN	5thL-UN	5thL-UN	5thL-UN	LB+M-P	Oldest leaf	MRFDL-WP	MRFDL-WP	MRFDL-WP	MRFDL	MRFDL	MRFDL
Age-Time	End of silk	FB start to SMFR	SMFR-Harvest	FFDL	MRFDL	8L stage	Pre-bulb	Bulbing	Harvest			
Number	8	12	12	12	12	10	12	12	12	12	12	12
						Analyses						
N %	2.2-2.7	4.5-6.0	4.0-5.5	4.3-6.0	4.0-6.0	4.3-5.0	4.4-5.0	3.4-5.5	2.9-3.5	2.0-3.5	3.1-5.5	4.0-5.0
P %	0.25-0.4	0.34-1.25	0.25-1.0	0.3-1.0	0.3-1.2	0.4-0.7	0.3-0.6	0.28-0.5	0.26-0.4	0.2-0.5	0.3-0.7	0.3-0.7
K %	1.4-2.5	3.9-5.0	3.5-4.5	3.1-5.5	3.5-5.0	5.0-6.0	3.9-4.8	3.0-4.5	1.8-2.8	2.5-4.5	2.0-4.0	3.3-4.5
Ca %	0.6-1.1	1.4-3.5	1.5-4.0	2.5-4.0	1.0-2.5	1.5-2.5	0.8-1.5	1.0-1.8	1.5-2.5	2.3-3.0	1.3-2.5	2.8-3.5
Mg %	0.2-0.5	0.3-1.0	0.3-1.2	0.35-1.0	0.3-1.0	0.25-0.5	0.15-0.25	0.23-0.3	0.25-0.35	0.25-0.5	0.25-0.7	0.3-0.5
S %	0.2-0.75	0.4-0.7	0.3-1.0	0.4-0.7			0.23-0.3					0.4-0.8
B ppm	8-25	25-60	30-100	30-100	25-75	25-75				25-60	30-100	30-75
Cu ppm	5-25	7-20	8-20	8-10	8-60	5-25				8-25	4-25	5-30
Fe ppm	50-300	50-300	50-300	50-300	50-300	40-150				50-200	60-300	50-300
Mn ppm	31-300	50-300	50-400	50-300	40-250	15-250				50-250	30-250	50-250
Mo ppm	0.9-10	0.8-3.3	0.8-3.3	0.8-3.3							0.1-0.15	
Zn ppm	20-150	25-100	25-200	25-200	20-250	30-250				25-200	30-250	25-250

TABLE 5.1. Vegetables (continued)

Plant	Leek	Lettuce, Boston	Lettuce, Boston	Lettuce, Cos type	Lettuce, Ithaca	Onion	Onion	Pea, English	Pea, Southern	Pepper Bell	Pepper Bell	Potato, Irish
Tissue	MRFDL	OLDL	WRL	WRL	WRL	TOPS-WHB	TOPS-WHB	LFT MRFDL	MRFDTR	MRFDL #1BL-1\3F	MRFDL F1\3	MRFDL
Age-Time		8LStage	Mature	Mature	Mature	1\3-1\2M	1\2-MAT	1STBL	EARLYBL			12"
Number	12	12	12	12	12	12	12	50	12	25	25	25
						—Analyses—						
N %	3.5-5.0	4.7-5.5	4.0-5.0	3.5-4.5	4.0-5.0	5.0-6.0	4.5-5.5	4.0-6.0	4.0-5.0	4.0-6.0	3.5-5.0	4.5-6.0
P %	0.3-0.6	0.5-1.0	0.4-0.6	0.45-0.8	0.4-0.6	0.35-0.5	0.3-0.5	0.3-0.8	0.3-0.6	0.35-1.0	0.22-0.7	0.29-0.5
K %	3.5-5.0	7.5-9.0	6.0-7.0	5.5-6.2	6.0-7.0	4.0-5.5	3.5-5.0	2.0-3.5	2.2-3.0	4.0-6.0	3.5-4.5	9.3-11.5
Ca %	0.6-0.9	2.0-3.0	2.3-3.5	2.0-2.8	2.3-3.5	1.0-2.0	1.5-2.2	1.2-2.0	2.0-3.0	1.0-2.5	1.3-2.8	0.75-1.0
Mg %	0.15-0.3	0.5-0.8	0.5-0.8	0.6-0.8	0.5-0.8	0.25-0.4	0.25-0.4	0.3-0.7	0.3-0.5	0.3-1.0	0.3-1.0	1.0-1.2
S %	0.5-1.0					0.5-1.0	0.5-1.0	0.3-0.5				
B ppm	25-75	25-50	25-60	25-60	25-60	22-60	25-75	25-60	25-80	25-75	25-75	25-50
Cu ppm	7-25	8-25	8-25	5-25	8-25	15-35	15-35	7-25	6-25	6-25	6-25	7-20
Fe ppm	50-250	50-100	50-100	40-100	50-100	60-300	60-300	50-300	50-100	60-300	60-300	50-100
Mn ppm	50-250	15-250	15-250	11-250	15-250	50-250	50-250	30-400	50-300	50-250	50-250	30-250
Zn ppm	20-50	25-250	25-250	20-250	25-250	25-100	25-100.	25-100	20-100	30-200	20-200	45-250

Plant	Potato, Irish	Potato, Sweet	Radicchio	Radish	Spinach	Spinach	Squash	Tomato, Trellis				
										PLOBTFC		
Tissue	MRFDL	MRFDL	WRL	MRFDL	MRFDL	MRFDL	B+M-P FMRFDL	MBL#1FCL	MBL#2FCL	MBL#3FCL	MBL#4FCL	MBL#5FCL
Age-Time	TU1\2MAT	1\2Grown	2\3Grown		30-50days	Mature						
Number	25	15	15	30	15	15	12	12	12	12	12	12
						—Analyses—						
N %	3.0-4.0	3.3-4.5	3.5-6.0	3.0-6.0	4.0-6.0	3.5-5.5	4.0-6.0	3.5-5.0	3.2-4.5	3.0-4.0	2.3-3.5	2.0-3.0
P %	0.25-0.4	0.23-0.5	0.25-0.6	0.3-0.7	0.3-0.6	0.25-0.5	0.3-1.0	0.7-1.3	0.5-1.2	0.4-1.0	0.25-1.0	0.2-0.8
K %	6.0-8.0	3.1-4.5	5.5-7.0	4.0-7.5	5.0-8.0	4.0-5.5	3.0-5.0	6.0-10.0	5.0-10.0	5.0-9.0	4.0-8.0	3.8-7.0
Ca %	1.5-2.5	0.7-1.2	1.0-2.0	3.0-4.5	0.7-1.2	0.8-1.5	1.2-2.5	1.4-2.2	1.5-2.4	1.5-2.4	1.5-2.5	1.5-2.5
Mg %	0.7-1.0	0.35-1.0	0.3-0.7	0.5-1.2	0.6-1.0	0.7-1.2	0.3-1.0	0.3-0.7	0.32-0.8	0.32-0.8	0.32-0.8	0.35-0.9
S %			0.3-0.5									
B ppm	40-70	25-75	25-60	25-125	25-60	25-60	25-75	25-75	25-75	25-75	25-75	25-75
Cu ppm	7-20		7-25	5-25	5-25	5-25	10-25	5-50	5-50	5-50	5-50	5-50
Fe ppm	40-100	40-100	75-500	50-200	60-200	60-200	50-200	60-300	60-300	60-300	60-300	60-300
Mn ppm	30-250	40-250	40-150	50-250	30-250	30-250	50-250	50-250	50-250	50-250	50-250	50-250
Zn ppm	30-250	20-50	20-75	19-250	25-100	25-100	20-200	20-250	20-250	20-250	20-250	20-250

TABLE 5.1. Vegetables (continued)

Plant	Tomato Trellis	Tomato, Field	Turnip	Water-cress	Water-melon	Water-melon
Tissue	PLOBTFC	COLADIINF	MRFDL	MRFDCLf	#5LFT-UN FLST-SNFR	#5LFT-UN SMFR-HAR
Age-Time	M8L#6FCL Midbloom	Midbloom				
Number	12	15	12	25	12	12
						--Analyses--
N %	1.8-2.5	4.0-6.0	3.5-5.0	4.2-6.0	4.0-5.5	4.0-5.00
P %	0.18-0.6	0.25-0.75	0.33-0.6	0.7-1.3	0.3-0.8	0.25-0.7
K %	3.5-6.0	2.9-5.0	3.5-0.6	4.0-8.0	4.0-5.0	3.5-4.5
Ca %	1.5-2.5	1.0-3.0	1.5-4.0	1.0-2.0	1.7-3.0	2.0-3.2
Mg %	0.35-0.9	0.4-0.6	0.3-1.0	0.25-0.5	0.5-0.8	0.3-0.8
S %		0.4-1.2				
B ppm	25-75	25-60	40-100	25-50	25-60	25-60
Cu ppm	5-50	5-20	6-25	6-15	6-20	6-20
Fe ppm	60-300	40-200	40-300	50-100	50-300	50-300
Mn ppm	50-250	40-250	40-250	50-250	50-250	50-250
Zn ppm	20-250	20-50	20-250	20-40	20-50	20-50

MISCELLANEOUS CROPS

Plant	Cocoa	Coffee	Coffee	Favabeans	Oil palm		Tea
Tissue	LPLP	L	#4L pairs from tip FLST MT trees	BL Top ML	Middle third portion #3 frond upper and 3 lower LF from	MR of 3 e 3 #17 frond	Third leaf from tip of young shoots
Age-time	Podfill 8-10:30AM	YNFB			Young trees	Mature trees	
Number	10\tree	50	100	50			100
				--------Analyses--------			
N %	2.0-2.5	2.5-3.5	2.3-3.0	4.8-5.0	2.8-3.0	2.7-3.0	3.8-4.8
P %	>0.18	0.15-0.35	0.12-0.2	0.3-0.42	0.19-0.25	0.18-0.25	0.19-0.25
K %	1.3-2.2	2.0-3.0	2.0-2.5	2.4-3.2	1.5-1.8	>1.3	1.8-2.0
Ca %	>0.40	1.0-2.2	1.0-2.5	0.5-0.75	0.3-0.5	>0.6	0.4-0.6
Mg %	>0.45	0.3-0.5	0.25-0.4	0.35-0.42	0.3-0.5		0.15-0.3
S %		0.25-0.5	0.01-0.2				0.10-0.3
B ppm	25-70	25-75	40-75			10-25	30-50
Cu ppm	8-12	10-50	10-25			5-8	
Fe ppm	60-200	90-300	70-125				500-1000
Mn ppm	50-300	50-300	50-200		150-200		
Mo ppm	1.0-2.5		0.1-0.5		0.5-1.0		Cl 30-50
Zn ppm	20-100	15-200	12-30		15-20		30-50

TABLE 5.1. (continued)

ABBREVIATIONS

B=blades
BBL=before bloom
B Head=before heading
BL=bloom
BLC=base of live crown
BPS=before pod start
CLF=center leaflets
COLADTINF=compound leaf adjacent to top inflorescence
CRL=crown leaves
CS=current season
CY=current year
DL=D leaf
EARANT=early anthesis
EB=head emerges from boot
EL=ear leaf
EP=early pegging

L=leaf
IS=initial silk
LB=leaf blade
LBF=leaf behind fruit
LBL=leaf below whorl
LFPR=leaflet pairs
LFT=leaflet
LP=large petiole
LPLP=lower part leaf pulvinus
M=midrib
MAT=mature
MB=mature berries
MBL=midbloom
MIDC=midway between cuttings
MIDL=midleaf section
MID-SUM=midsummer
MLF=middle leaflets

NFLU=nonflushing
NFL=nonflowering
NFT=nonfruiting terminals
OBC=opposite bunch cluster
OBFC=opposite flower cluster
OL-1"B&TN=outside leaves minus 1 inch base and node
OLDL=oldest leaf
P=petiole
PI=plant initiation
PLOBTFC=petioles of leaves opposite or below top flower cluster
POBL=postbloom
RMGS=recently mature growth shoots
S=stems
SMFR=small fruit
ST=start

F=final
FB=full bloom
FCL=flower cluster
FEL=fully expanded leaf
FFDL=first fully developed leaf
FFDS+L=1st fully expanded stem + leaves
FFL=first flower
FFMN=first fully mature needles
FL=flowering
FLBST=flower bud start
FLST=flower start
FR=frond
HAR=harvest
INFST=inflorescence start

MPTG=midportion terminal growth
MRDCL=most recent developed
compound leaf
MRDL=most recent developed leaf
MRFD=most recent fully developed
MRFDTR=most recent developed
trifoliate leaf
MRFM=most recent fully mature
MS=midshoot
MSPF=midshoot previous flush
MT=mature tree
MTIL=maximum tillering
N=needles
NF=nonfruiting

T=tassel
Ti=tips
TMLB=topmost lateral branch
TRL=trifoliate leaf
TU=tubers
UN=unfurled
UP=upper part of plant
US=uppermost shoots
VEG=vegetative
VEGSH=vegetative shoot
WHB=white base
WP=white portions
WRL=wrapper leaf
YNFB=young nonfruiting branches

Source: A compilation of many sources, including the results from fifty years of the author's analytical data. Listed values are associated with satisfactory yields. Values below limits are associated with reduced yields, but those above upper limits may not be indicative of toxicity although response to additional fertilizer cannot be expected. Safe values for copper may be appreciably higher than upper values listed if copper fungicides have been used.

TABLE 5.2. Interpretive guide for tissue testing*

Plant	Alfalfa	Asparagus	Bean, bush\snap		Broccoli		Brussels sprouts		Cantaloupe		
Tissue	Tops MS+L	10 cm tip new fern	Petiole 4th leaf from tip		Midrib of young mature leaf		Midrib of young mature leaf		Petiole 6th leaf from tip		
Age-time	1\10 bloom	Mid-growth	Mid-growth	Early bloom	Mid-growth	1st buds	Mid-growth	Late growth	Early growth	Early fruit	First harvest
Number	40	15	25	25	12	12	12	12	25	25	25
					—Analyses—						
NO3-N ppm Def.	<100	<2000	<1000	<5000	<7000	<5000	<7000	<5000	<8000	<5000	<2000
Suff.	>500	>3000	>1500	>7000	>9000	>7000	>9000	>7000	>12000	>8000	>3000
PO4-P ppm Def.	<500	<1000	<800	<2500	<2500	<2500	<2000	<2000	<2000	<1500	<1000
Suff.	>800	>2000	>1500	>4000	>4000	>4000	>3500	>3000	>3000	>2500	>2000
K % Def.	<1	<3	<2	<2	<3	<2	<3	<2	<4	<3	<2
Suff.	>3	>5	>4	>4	>5	>4	>5	>4	>6	>5	>4

Plant	Cabbage	Cabbage Chinese	Cantaloupe			Carrot	Cauliflower	Celery		Clover Ladino	Corn sweet
Tissue	Midrib of WRL	Midrib of WRL	Blade of 6th leaf from tip			Petiole of FFDL	Midrib of FFDL	Petiole from most recent mature leaf		Petiole of FFDL	Midrib of FFDL
Age-time	Heading	Heading	Early growth	Early fruit	1st mat. fruit	Mid-growth	Buttoning	Midgrowth	Near mature		Mid-season
Number	15	15	12	12	12	25	12	12	12	25	25
					—Analyses—						
NO3-N ppm Def.	5000	8000	<2000	<1000	<500	<5000	<5000	<5000	<4000		<500
NO3-N ppm Suff.	7000	10000	>3000	>1500	>800		>7000	>7000	>6000		
PO4-P ppm Def.	2500	2000	<1500	<1300	<1000	<2000	<2500[xx]	<2000	<2000	<600	<500
PO4-P ppm Suff.	3500	3000	>2300	>2300	>1500	>3000	>3500[xxx]	>5000	>3000	600-1200	
K % Def.	2	4	<1	<1	<1	<4	<2	<4	<3	<0.8	2
K % Suff.	4	7	>2.5	>2	>1.8	>6	>4	>6	>5	0.8-1.75	4

TABLE 5.2. (continued)

Plant	Cucumber pickling	Eggplant	Lettuce	Onion	Pepper chili	Pepper sweet	Potato Irish	Potato Irish	Potato Irish	Potato sweet
Tissue	Petiole of FFDL	Petiole of FFDL	Petiole of FFDL	Tallest leaf	Petiole of FFDL	Petiole of FFDL	Petiole of FFDL	Petiole of FFDL	Petiole of FFDL	Petiole of FFDL
Age-time	Mid-season	First harvest	Mid-season	Mid-season	Mid-season	Mid-season	Early	Midseason	Late	Mid-season
Number	25	15	15	12	50	50	25	25	25	25
					Analyses					
NO_3-N ppm Def.	5000	5000	<4000		5000	8000	<8000	<6000	<4000	<1500
NO_3-N ppm Suff.		7500	>8000				>11000	>8000	>5000	
PO_4-P ppm Def.	<1500	2000	<2000	<1000	2000	2000	<1200	<900	<600	1000
PO_4-P ppm Suff.	3000	3000	>4000	>2000	2000	2000	>2000	>1500	>1000	
% Def.	3	4	2	<2.5	4	4	<9	<7	<4	3
% Suff.		7	4	>4.0		>11	>9	>6		

Plant	Rice				Spinach	Sugarbeet			Tomato (canning)		
Tissue	short grain varieties — Recently matured Y leaves				Petiole of FFOL	Petiole	Petiole of FFOL xxxx	Blade of FFOL xxxx	Petiole of 4th leaf from growing tip		
Age-time	Mid til-lering	Max til-lering	Panicle formation	Flagging	Mid-growth	Seedling			Early bloom	One inch fruit	First color
Number	50	50	50	50	12	25	25	12	12	12	12
— Analyses —											
NO3-N ppm — Def.					<4000	<1500	<1000		8000	6000	2000
NO3-N ppm — Suff.					>6000	1600-5000	1000-35000		12000	10000	4000
PO4-P ppm — Def.	<1000xx	<1000xx	<800xx	<800xx	<2000xx		<750	<700	1000	1000	800
PO4-P ppm — Suff.	1000-1800	1000-1800	800-1000	800-1800	>3000xx		750-4000	1000-8000	2500	2500	1800
K % — Def.	<1.4xx	<1.2xx	<1.0xx	<1.0xx	<2	<1	<1	<1	3	2	1
K % — Suff.				>4	>10	1-6	1-11	1-6	6	4	3

TABLE 5.2. (continued)

Plant	........Tomato........		........Tomato........ cherry market		Watermelon
Tissue	Petiole 4th leaf		....Petiole.... 4th leaf from growing tip		Petiole of FFDL
Time-age	Early fruit set	Early fruit set	One inch fruit	Full ripe fruit	Mid-season
Number Analysis	12	12	12	12	12
NO3-N ppm Def.	8000	10000	8000	5000	5000
Suff.	10000	14000	12000	9000	9000
PO4-P ppm Def.	2000	2500	2500	1500	1500
Suff.	3000	3500	3500	2500	2500
K % Def.	5	4	4	3	3
Suff.	7	6	6	5	5

*Two % acetic acid solution NO3-N and PO4-P, and total K (dry weight basis).

** Low values

*** High values

**** Sodium affects K content of sugarbeet leaves making it mandatory to use blades for K analyses if petioles contain <1.5 % Na

Abbreviations: FFDL = First fully developed leaf

          mat = mature

Source: *Soil and plant tissue testing in California.* 1976. (ed.) H. N. Reisenour. Univ. of California Div. of Agric. Sci. Bull. 1879. With modifications from Ulrich, A. and F. Jackson. 1990. *Plant analyses as an aid in fertilizing sugarbeet.* In *Soil testing and plant analyses.* Third edition, (ed.) R. L. Westerman. Soil Sci. Soc. of America, Inc., Madison, WI 53711 and Fulmer, F. S. and M. E. McCollum. 1964. *Ask the leaf.* Better Crops with Plant Food, May-June.

TABLE 5.3. Adequate or critical values of sap NO_3-N and K in several vegetable crops

Crop	Sampling time	Plant part*	NO_3-N ppm	K ppm
Broccoli and collards	6th leaf stage	Petiole	800-000	
Collards	Prior to harvest	Petiole	500-800	
	1st harvest	Petiole	300-500	
Cantaloupe	1st blossom	Petiole	1000-1200	
	1st fruit 2 inches	Petiole	800-1000	
	1st harvest	Petiole	700-800	
Cucumber	1st blossoms	Petiole	800-1000	
	1st fruit 3 inches	Petiole	600-800	
	1st harvest	Petiole	400-600	
Eggplant	1st fruit 2 inches	Petiole	1200-1600	4500-5000
	1st harvest	Petiole	1000-1200	4000-4500
	Mid harvest	Petiole	800-1000	3500-4000
Lettuce	Harvest	Entire	2000**	

149

TABLE 5.3. (continued)

Crop	Sampling time	Plant part*	NO₃-N ppm	K ppm
Pepper	1st flower buds	Petiole	1400-1600	3200-3500
	1st open flowers	Petiole	1400-1600	3000-3200
	Fruit half-grown	Petiole	1200-1400	3000-3200
	1st harvest	Petiole	800-1000	2400-3000
Potato	18 days growth	Terminal Petiole	2000**	
	8 inches tall	Petiole	1200-1400	4500-5000
	1st open flowers	Petiole	1000-1400	4500-5000
	50% flowering	Petiole	1000-1200	4000-4500
	100% flowering	Petiole	900-1200	3500-4000
	Tops falling	Petiole	600-900	2500-3000
	Tops down	Petiole	200-800	<2000
Spinach	Harvest	Petiole	1700**	
Summer squash	1st blossoms	Petiole	900-1000	
	42 days	Petiole	1000**	
	from seeding			
	1st harvest	Petiole	800-900	
Strawberries***	30 days	Petiole	800-900	3000-3500
	60 days	Petiole	600-800	3000-3500
	90 days	Petiole	600-800	2500-3000
	120 days	Petiole	300-500	2000-2500
	150 days	Petiole	200-500	1800-2500
	180 days	Petiole	200-500	1500-2000

Crop	Sampling time	Plant part*	NO$_3$-N ppm	K ppm
Tomato (field)	1st buds	Petiole	1000-1200	3500-4000
	1st open flowers	Petiole	600-800	3500-4000
	1st fruit 1 inch	Petiole	400-600	3000-3500
	1st fruit 2 inches	Petiole	400-600	3000-3500
	1st harvest	Petiole	300-500	2500-3000
	Main harvest	Petiole	300-400	2000-2500
Tomato (greenhouse)	Transplant to 2nd cluster	Petiole	1000-1200	4500-5000
	2nd cluster to 5th cluster	Petiole	800-1000	4000-5000
Watermelon	Vines 6 inches	Petiole	1200-1500	4000-5000
	1st fruit set	Petiole	1200-1500	
	1st fruit 2 inches	Petiole	1000-1200	4000-5000
	Fruit half mature	Petiole	800-1000	3500-4000
	1st harvest	Petiole	600-800	3000-3500

* Unless otherwise stated, sap is derived from petiole of first fully developed leaf.

**Critical values.

***Annual hill strawberries planted in Florida.

SOURCE: A compilation of data from Hochmuth, G. 1994. Efficiency ranges for nitrate-nitrogen and potassium for vegetable petiole sap quick tests. *HortTechnology*, 4:218-222; Hochmuth, G. and E. Albregts. 1994. *Strawberries*. Fertilization of Strawberries in Florida. University of Florida Coop. Ext. Serv. Circular 1141; Maynard, D. N., A. V. Barker, P. L. Minotti, and N. H. Peck. 1976. Nitrate accumulation in vegetables. *Advances in Agronomy* 28:71-118; Vitosh, M. L. and G. H. Silva. 1994. A rapid petiole sap nitrate-nitrogen test for potatoes. *Commun. Soil Sci. Plant Anal.*, 25:183-190.

Chapter 6

Maturity Indices

There are several procedures which help a grower decide the proper timing of the harvest in order to provide a mature product. Maturity may be, but in most cases is not, synonymous with ripeness. According to M. S. Reid of the University of California at Davis, maturity is "that stage at which a commodity has reached a sufficient stage of development that after harvesting and postharvest handling (including ripening, where required) its quality will be at least the minimum acceptable to the ultimate consumer" (Reid, 1992).

Methods that evaluate maturity can reduce losses of harvested commodities. A large number of indices for determining maturity of some fruits and vegetables are presented in Table 6.1. Some of the more common methods of measurement are outlined below.

SOLUBLE SOLIDS

A number of fruits undergo substantial increase in soluble solids as they approach maturity. Some minimum legal standards have been set for several fruits based on soluble solids, soluble solids/acid ratios and/or soluble solids and firmness. (See Table 6.2.)

Soluble solids are easily measured in the field by means of a hand refractometer. A drop or two of juice squeezed from the sample is placed on the prism. The sample is spread with a plastic or unfinished wood stirring rod to cover the prism. The cover is closed and readings are obtained by pointing the instrument to a light source and noting where the dividing line between the light and dark fields crosses the

TABLE 6.1. Indices used for determining maturity of fruits and vegetables

Index	Method of determination	Examples
Elapsed days from full bloom	Computation	Apple, pear
Mean heat units	Computation from weather data	Pea, apple, sweet corn
Development of abscission layer	Visual or force of separation	Apple, feijoa, some melons
Surface morphology and structure	Cuticle formation on grape, tomato Netting of some melons Gloss of some fruits	Visual
Size	Measuring devices, weight	All fruits and many vegetables
Specific gravity	Density gradient solutions, flotation techniques, vol/wt	Cherry, potato watermelon
Shape	Dimension ration charts	Angularity of banana fingers Full cheek of mango Compactness of broccoli and cauliflower
Solidity	Feel, bulk, density, y-rays, X-rays	Brussels sprouts, cabbage, lettuce
Textural		
Firmness	Firmness testers,	Apple, pear, stone
Tenderness	deformation	fruits
Roughness	Tenderometer	English peas
	Texturometer, fibrometer	Celery
	Chemical tests for polysaccharides	
Color, external	Light reflectance Visual color charts	All fruits and some vegetables
Color, internal	Light transmittance, delayed light emission	Jellylike material in tomato, flesh color of some fruits

Index	Method of determination	Examples
Composition		
Total solids	Dry weight	Avocado, kiwi fruit
Starch content	Starch-iodine test	Apple, banana, pear
Sugar content	Refractometer, chemical tests	Apple, grape, melons, stone fruits
Acid content, sugar: acid ratio	Sugar as above; acid Titration	Citrus, kiwi, melons, Papaya, pomegranate
Juice content	Extraction, chem. tests	Citrus
Oil content	Extraction, chem. tests	Avocado
Tannin content	Chemical tests	Date, persimmon
Ethylene content	Gas chromatography	Apple, pear

Source: A compilation of data from three tables presented by Read, M. S. 1992. Maturity and maturity indices. In *Postharvest Technology of Horticultural Crops.* A. A. Kader (ed.). Publication No. 3311. Division of Agriculture and Natural Resources. University of California, Oakland, CA 94608-1239

scale. Most instruments read in Brix units or percent sugar. A typical hand-held refractometer is depicted in Figure 6.1.

SOLUBLE SOLIDS/ACID RATIOS

As several fruits ripen, acid concentration falls while soluble solids (mainly sugars) rise. The ratio, determined by simple division, is used as an index of maturity, but there is a sliding scale for different citrus fruits. Higher amounts of acid are allowed with higher Brix (lower ratio) than with lower Brix, but if sugars are low, ratio requirements are increased. This variation of ratio to compensate for varying solids allows for different standards for the different fruits at varying times of the year. For example, the 9:1 ratio of solids to acid permissible for oranges with percentage solids of at least 11.0 percent on August 1 gradually changes to a ratio of 10.15:1 for minimum solids of 8.7 percent by November 1. The

FIGURE 6.1. Hand refractometer

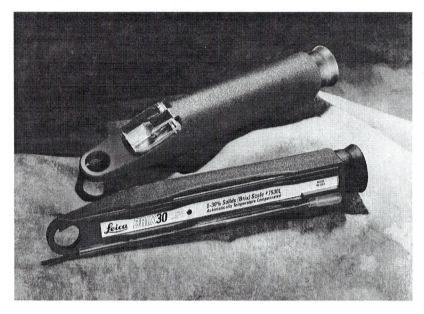

Hand refractometers enable very rapid determination of sugars or solids in the field. Photo courtesy of Leica Products Division, Leica Inc., Buffalo, NY 14240-0123.

acceptable ratio of solids to acid of 8.6:1 for fresh tangelos with a minimum of 11.8 percent solids on August 1 gradually changes to 10:1 for minimum solids of 9 percent by November 15. (The minimum quality requirements for various citrus fruits in Florida can be obtained from the Florida Division of Fruit and Vegetable Inspection, P.O. Box 1072, Winter Haven, FL 33882-1072.)

Soluble solids/acid ratios in Florida are run on the juice of oranges, grapefruits, tangerines, temples, and tangelos. The juice is extracted by hand (reamer or orange burr) or by mechanical extractor and then filtered through a double layer of cheesecloth, a wire strainer, or colander. Sufficient pressure is applied to squeeze out the juice while leaving the juice sacs, pulp, seeds, and "rag" behind.

TABLE 6.2. Legal limits of fruit maturity based on minimum soluble solids, soluble solids:acid ratios, and firmness

Fruit	Agency	Soluble solids %	Soluble solids–acid ratio	Firmness
Apple	Calif			
Red Delicious		11.0	–	
	18			
Golden Delicious		12.0	–	18
Jonathan		12.0	–	19
Rome		12.5	–	21
Newton Pippin		11.0	–	23
McIntosh		11.5	–	19
Gravenstein		10.5	–	–
Cantaloupe	Calif	>8	–	–
	USDA	>9	–	–
Cherry, sweet	Calif	14-16*	–	–
Citrus			–	–
Grapefruit	Calif	–	5.5 or more	–
Grapefruit	Calif**	–	6 or more	–
Orange	Calif	–	8 or more	–
Tangerine or Mandarin	Calif	–	6.5 or higher	–
Grape, Table	States***			
Muscat		17.5	–	–
Blanche, Cardinal		15.5	–	–
Emperor, Olivette				
Red Malaga,				
Ribier, Rosh Baba				
All others		16.5	–	–
Grape processing				
American type	USDA	>15.5	–	–
Grape juice	USDA	>16-18	–	–
	Calif	14-17.5****	20 or more	–
Honeydew	Calif	>10	–	–
Kiwi fruit	Calif	>6.5	–	–

*   Based on cultivar.
**   For desert areas.
***   For states other than Arizona or California.
****   Depending on cultivar and production area.
Source: Based on data of Kader, A. A. Standardization and inspection of fresh fruits and vegetables. 1992. In *Postharvest Technology of Horticultural Crops*. A. A. Kader (ed.). Publication No. 3311. Division of Agriculture and Natural Resources, University of California.

Soluble solids of the juice can be measured by Brix hydrometer at a temperature of 68°F. Different temperatures can be used but corrections must be made accordingly. (See Table 6.3.) Acid is determined by titrating the juice with a standard sodium hydroxide solution (0.3125 N) using a direct reading burette to give total acid (really titratable acid). Titration is carried out with three to four drops of phenolphthalein solution using constant agitation until a pink (but not red) color is obtained. An ordinary burette can be used, but a conversion table is needed to provide percent anhydrous citric acid. In a nonofficial method, titration with a stronger sodium hydroxide solution (0.4063 N) with an ordinary burette allows for each cubic centimeter of alkali used to equal 0.1 percent citric acid.

STARCH

The amount of starch and its distribution change in some fruits as they ripen. The disappearance of starch in apples, pears, and bananas as they ripen can be used as a maturity index. The exposed flesh of apples or pears cut in half, or of an unpeeled banana with a fresh cross-section cut across its center and a cut surface on its blossom end, are treated with a solution of potassium iodide (1 percent) plus iodine (0.1 percent) in distilled water or alcohol. The presence of starch is noted by the amount of dark blue-black color produced as iodine reacts with the starch.

FIRMNESS

Many fruits soften as they ripen and the degree of firmness as measured with a firmness tester can be used as an index of their maturity. Combining firmness with soluble solids helps evaluate maturity of apples. (See Table 6.4.) Other fruits for which firmness can be used as an index of maturity are: grapefruit, tangerine, cranberry, pear, and pineapple. Using firmness along with the starch-iodine index can be useful in determining the suitability of apples for long-term controlled atmosphere (CA) storage. (See Table 6.4.)

Some vegetables tend to become tough or fibrous as they overmature. The force required to push a standard probe of a firmness

TABLE 6.3. Temperature corrections for degrees Brix to standard temperature of 20° C.

Temperature °C	Correction Factor °Brix	Temperature °C	Correction Factor °Brix
10.0	−0.45	25.0	+0.25
11.0	−0.40	25.5	+0.30
12.0	−0.40	26.0	+0.35
13.0	−0.35	26.5	+0.35
14.0	−0.30	27.0	+0.40
15.0	−0.30	27.5	+0.40
15.5	−0.25	28.0	+0.45
16.0	−0.25	28.5	+0.50
16.5	−0.20	29.0	+0.55
17.0	−0.15	29.5	+0.55
17.5	−0.10	30.0	+0.60
18.0	−0.10	30.5	+0.65
18.5	−0.05	31.0	+0.65
19.0	−0.05	31.5	+0.70
19.5	0.00	32.0	+0.75
20.0	0.00	32.5	+0.75
20.5	+0.05	33.0	+0.80
21.0	+0.05	33.5	+0.85
21.5	+0.10	34.0	+0.90
22.0	+0.10	34.5	+0.90
22.5	+0.15	35.0	+0.95
23.0	+0.15	35.5	+1.00
23.5	+0.20	36.0	+1.05
24.0	+0.20	36.5	+1.10
24.5	+0.25	37.0	+1.10
		37.5	+1.15

Note: For certain hydrometers calibrated to 17.5° C, shift the values in this table by 2.5° C so that the zero correction factor corresponds to 17.5° C.

Source: Wardowski, W., J. Soule, J. Whigham, and W. Grierson. 1991. *Florida Citrus Quality Tests*. SP 99. Florida Cooperative Extension Service/IFAS, Gainesville, FL.

TABLE 6.4. Starch-iodine indices and firmness values associated with long-term controlled atmosphere (CA) storage of apple varieties grown in Michigan

Cultivar	Starch-iodine index*	Firmness (lb)**
Empire	4.5-5.5	16-17
Jonagold	3-5.5	15-16
Jonathan	3-5.5	15.5-16
McIntosh	4-6	4.5-15
	For export	
Golden Delicious	3-5	15-17
Red Delicious	2.5-4	17-18
Idared	3-4	14.5-16
Mutsu	3-4.5	16-17.5
Law Rome	3-5	18-19

*Based on Ontario and North Carolina starch index charts. The Ontario chart is described by Chu, C. L. 1988. Starch-iodine test for determining maturity and harvest dates of Empire, Idared, and Spartan apples. Ont. Min. of Agric. and Food Bul. 88-090 and K. L. Priest and E. C. Lougheed. 1988. Evaluating maturity of McIntosh and Red Delicious apples. Ont. Min. of Agric. and Food Bull 88-117; the North Carolina chart by M. E. Salveit and S. A. Hale. 1982. Determining the maturity of North Carolina apples: The starch-iodine staining technique. North Carolina Agr. Ext. Serv.
**As measured by a Magness-Taylor penetrometer.

Source: Beaudry, R., P. Schwallier, and M. Lennington. 1993. Apple maturity prediction: An extension tool to aid fruit storage decisions. *HorTechnology* 3(2); 233-239.

tester through cabbage, brussels sprouts, and lettuce is indicative of their maturity. Also, tenderometer ratings of shelled English peas can be used to determine ideal harvesting time, since readings increase markedly as peas overmature.

JUICE CONTENT

The amount of juice increases in several fruits as they mature. The amount of juice can be used as an indicator for harvesting lemons. The USDA lists 28-30 percent (depending upon grade) as a minimal juice content (by volume), and California has a minimum value of 30 percent. Florida lists 30 percent juice by volume for the U.S. #1, U.S. Combination, and US #2 grades, but only 28 percent by volume for export grades of lemons. A minimum of 42 percent juice is required for limes.

COLOR

The color of many commodities change as they ripen, permitting the use of color to indicate the time to initiate harvesting. California uses color as one of the indices to measure maturity, and the minimum stipulations for several fruits are as follows: (1) the entire surface of sweet cherries have at least a solid red color; (2) more than 2/3 of grapefruit have a yellow color; (3) orange color on at least 25 percent of orange fruits; (4) yellow, orange, or red color on at least 75 percent of mandarin or tangerine fruits; and (5) 2/3 of the surface of strawberry fruits have a pink or red color. The colors specified as maturity indices for fruits and vegetables by California and several other states, as well as the USDA, often are given in precise color values of standard color charts (CDFA or Munsell). Measuring maturity is a matter of comparing commodity color with that on the chart. Use of a color analyzer can speed up the process and improve the accuracy of identifying color, but may not be acceptable legally. (See Chapter 1 for a description of an instrument used for color analysis.)

DRY WEIGHT

The dry weight of avocados is related to the amount of oil as these fruits ripen. Oil content of 12-16 percent, corresponding to dry weights of 22-26 percent, are associated with satisfactory taste scores.

The dry weight of sliced avocado can be determined by drying a 100-gram sample at a temperature of 170°F for a 24-hour period and reweighing the sample immediately after it is cooled.

Additional Readings

Seed Evaluation

Cardwell, V. B. 1984. Seed germination and crop production. In *Physiological Basis of Crop Growth and Development*. M. B. Tesar (ed.). American Society of Agronomy, Crop Science Society of America, Madison, WI 53711.

Maxon, S. (ed.). 1993. Rules for testing seeds. Association of Official Seed Analysts. *Journal of Seed Technology* 16(3):1-113.

Plant Analysis

Plant Tissue and Sap Analysis Manual. 1990. Hach Company, Loveland, CO 80539.

Chapman, H. D. (ed.). 1966. *Diagnostic Criteria for Plants and Soil*. Division of Agricultural Sciences, University of California. Descriptions and photographs of nutritional disorders and with considerable interpretative data of plant analysis.

Hochmuth, G. 1994. Efficiency ranges for nitrate nitrogen and potassium for vegetable petiole sap quick tests. *HortTechnology* 4:218-222.

Jones, J. B., Jr., B. Wolf, and H. A. Mills. 1991. *Plant Analysis Handbook*. Micro-Macro Publishing, Inc., Athens, GA 30607. A guide for sampling, preparation, analysis and interpretation of plant analysis.

Reuter, D. J. and J. B. Robinson (eds.). 1986. Plant Analysis: An Interpretation Manual. Inkata Press Pty Ltd., Victoria, Australia.

Scaife, M. A. and R. G. Bray. 1977. Quick sap tests for improved control of crop nutrient status. *ADAS Q. Rev.* 27:137-145.

Westerman, K. L. (ed.). 1990. *Soil Testing and Plant Analysis*. Third edition. Soil Science Society of America, Madison, WI 53711.

Maturity Indices

Beaudry, R., P. Schwallier, and M. Lennington. 1993. Apple maturity prediction: An extension tool to aid fruit storage decisions. *HortTechnology* (Apr/June) 3(2):233-239.

Kader, A. A. 1992. Standardization and inspection of fresh fruits and vegetables. In *Postharvest Technology of Horticultural Crops*. A. A. Kader (ed.). Division of Agriculture and Natural Resources, Publication No. 3311. University of California, Oakland, CA 94608-1239.

Reid, M. S. 1992. Maturation and maturity indices. In *Postharvest Technology of Horticultural Crops*. A. A. Kader (ed.). Division of Agriculture and Natural Resources, Publication No. 3311. University of California, Oakland, CA 94608-1239.

Wardowski, W., J. Soule, J. Wingham, and W. Grierson. 1991. *Florida Citrus Quality Tests*. SP 99. Florida Cooperative Extension Service/IFAS, University of Florida, Gainesville, FL.

SECTION III:
WATER DIAGNOSTICS

Diagnostic procedures can determine whether a particular source of water is suitable for irrigation and when it needs to be applied. If not suitable for irrigation in its natural state, diagnostic procedures often can suggest treatments to modify the water in order to minimize its harmful effects. There are also procedures for measuring the amount of water applied and determining the uniformity of application.

Chapter 7

Diagnosing the Need for Water

CONSIDERING IRRIGATION

Most cultivated plants require a continuous supply of water to produce a crop. Except in the seed stage, water is the major portion of a plant's composition, and is necessary for all vital functions.

There are few places on earth where cultivated plants do not require some irrigation to produce maximum economic yields (MEY). But in many regions, water is already scarce and becoming more difficult to obtain as population pressures build. Even in regions receiving excesses of water, there are times when irrigation can provide extra returns, but the grower must know precisely when these periods occur. Irrigation applied when there is a sufficient amount of water is not only a waste of a vital resource, but is an unneeded expense that can lead to lowered crop quality and yield. Also, adding excess water often leads to the leaching of nutrients that can contaminate ground waters.

Not all soil water is available to plants. The water may be present in three different forms that provide varying availability to the plant: (1) *hygroscopic*, which is tightly held by soil particles and of no value, (2) *capillary*, which is held by surface tension in pore spaces between soil particles, and is the principal source of water for the plant, and (3) *gravitational* or free water that moves downward with gravity. After saturating the soil, free or excess water will percolate downward with gravity leaving the soil at *field capacity* (maximum amount of water held by the soil after wetting and the excess is drained). Plants can use only the water that exists in a range of what is held at field capacity and that which is still present at the *wilting point* (level of moisture at which plants wilt permanently. All hygroscopic water and part of the capillary water held by the soil at tensions greater than 15 atmospheres is unavailable to plants.

Not all of the water held between field capacity and the wilting point is equally available. Plants tend to readily absorb the weakly held water near field capacity but find it increasingly difficult to remove the water as the amount decreases and approaches the wilting point. There is some water near or at the wilting point, but plants generally cannot make use of it. Only about 75 percent of the total available moisture is readily available for most plants.

The reduction of availability as the water decreases is due to the increasing tension of the water as it is held by the soil. The tension, resulting from capillary and adhesive forces, is zero at the point of field saturation, rises to about 1/3 atmosphere at field capacity, and to 15 atmospheres at the wilting point. In air-dried soils, the water is held by forces equivalent to about 1000 atmospheres.

The tension at which yields are adversely affected (critical point) varies with different crops. The yields of broccoli, celery, endive, lettuce, and strawberry usually decline if the available water falls below 70 percent, but yields of beet, rhubarb, sorghum, or watermelon are not lowered until the readily available water falls below 20 percent of the total.

There will be a response to irrigation at most stages of growth if the plant is under severe stress from lack of water, but response at the critical periods can be obtained even if stress is mild. The critical point varies with different stages of growth. For many plants, the period of seed germination and seedling development is critical. Relatively small reductions in the available water at flowering or the period of fruit development and enlargement can also seriously limit yields of many plants. Some critical periods for different crops are given in Table 7.1.

Soil Moisture Storage

The amount of water that a soil can hold, or its moisture holding capacity (MHC), affects irrigation scheduling. The MHC varies with the amounts of OM and clay content and can be estimated from data for the different soil classes or calculated from actual measurements. Approximate values, along with useful data on infiltration and other properties for the different classes, are given in

TABLE 7.1. Critical periods of moisture stress

Crop	Stress period*
Alfalfa	Soon after cutting and flower start for seed development
Apricots	Bud development and flowering
Barley	Early boot stage, tillering, and soft dough
Broccoli	Head formation and enlargement
Cabbage	Head formation and enlargement
Cantaloupe	Flowering to early fruit development
Citrus	Flowering to early fruit development
Corn	Silking and tasseling to soft dough
Olive	Just prior to flower start and fruit enlargement
Peach	Final swelling of fruit
Peanut	Flowering and seed development
Potato	Stolonization and tuber initiation
Rice	Head development
Safflower	Seed filling
Sorghum	Boot to soft dough
Sugar beet	A few weeks after emergence
Sugarcane	Tillering and stem elongation
Tobacco	Knee high to flower start
Tomato	Flowering and period of rapid fruit enlargement
Watermelon	Flowering to harvest
Wheat	Booting, flowering, and early grain formation

*Seed germination is a stress period for most plants.

Source: Compilation from many different sources and the author's observations over a period of 50 years.

Table 7.2. Approximate amounts of water needed per acre for different soil textural classes and root zones to reach field capacity are given in Table 7.3.

Available Water from Field Capacity and Wilting Point Measurements

There are two approaches for determining the amounts of water held between field capacity and that at the wilting point. In the first, which is strictly a laboratory procedure, field capacity is determined

TABLE 7.2. Moisture holding capacities and other important physical properties* of different textured soils

Soil texture	Infilt. perm.	Total pore space	Field capacity	Perm. wilting point	Total Available moisture	
					Weight	Volume
	in/hr	%	%	%	%	%
Sandy	2 (1-10)	38 (32-42)	6 (6-12)	4 (2-6)	5 (4-6)	8 (7-10)
Sandy loam	1 (0.5-3.0)	43 (40-47)	14 (10-18)	6 (4-8)	8 (6-10)	12 (9-15)
Loam	0.52 (0.3-0.8)	47 (43-49)	22 (18-26)	10 (8-12)	12 (10-14)	17 (14-20)
Clay loam	0.32 (0.1-0.6)	49 (47-51)	27 (23-31)	13 (11-15)	14 (12-16)	19 (16-22)
Silty loam	0.10 (0.01-0.2)	51 (49-53)	31 (27-35)	15 (13-17)	16 (14-18)	21 (18-23)
Clay	0.2 (.004-0.4)	53 (51-55)	35 (31-39)	17 (15-19)	18 (16-20)	23 (20-25)

*Normal values are shown in parentheses. The lower values in each category are more appropriate for soils low in OM and the higher values for high organic soils.

Source: Hansen, V. E., O. W. Israelson, and G. E. Stringham. 1979. *Irrigation Practices and Principles.* Salt Lake City, Ut. Reprinted by permission of John Wiley & Sons, Inc.

TABLE 7.3. Approximate amounts* of water needed per acre for different soil textural classes and root zones to reach field capacity

Depth of root zone ('')	Sands		Loams		Clays	
	inches	gal/a	inches	gal/a	inches	gal/a
9	0.25-0.5	6500-13500	0.5-0.75	13500-20000	0.75-1.0	20000-27000
18	0.5-1.0	13500-27000	1.0-1.5	27000-40500	1.5-2.0	40500-55000
36	1.0-2.0	27000-54500	2.0-2.5	54500-67500	3.0-4.0	81000-110000

*Based on irrigation applied when one-half of available water in the effective root zone has been used. The lower values are for soils low in organic matter.

Source: Cook, R. L., J. R. Davis, and H. E. Kidder. *Fertilizing Through Irrigation Water.* Michigan State University Extension Bull. 324.

by measuring the moisture remaining in a sample of previously saturated soil that is held in a special apparatus at a tension of 5 pounds per square inch (psi) or 1/3 atmosphere. The wilting point moisture is determined by measuring the amount of moisture of the sample remaining after the soil is held in the apparatus at 15 atmospheres tension or 225 psi. In the second method, which is primarily a field procedure, field capacity is determined by heavily irrigating a field soil, allowing it to drain for a couple of days while it is covered to reduce evaporation, and then measuring the water held by gravimetric means (see Gravimetric section in this chapter). This second procedure requires a great deal more time but uses simple equipment for making the measurements. The wilting point moisture is also determined gravimetrically. Samples are collected from dwarf sunflowers grown in a series of cans in moist but not rewetted soil. Samples are taken when growth stops or permanent wilting takes place. Permanent wilting is indicated by an inability of leaves to recover their shape when plants are placed in a humid atmosphere.

IRRIGATION SCHEDULING

Once the storage capacity of the soil and the critical point are known, it is only necessary to estimate the amount of existing moisture to determine whether irrigation is needed. Some of the methods used for estimating existing soil moisture or its adequacy are outlined below.

Feel and Appearance

The simplest means of evaluating the need for irrigation employs plant appearance and feel of the soil. Wilting of the plant is used as an indicator for water need. The amount of water in the soil is estimated by the appearance and feel of a sample as it is compressed by hand and rubbed between the thumb and forefinger.

Although simple, the method has its limitations. It takes some time to train an operator to skillfully evaluate the soil moisture status from appearance and feel of the soil. Substantial yield may be lost by the time some plant symptoms appear. On the plus side, a great deal of useful knowledge about soil compaction and effective root zone can be gained by a skilled operator as he makes the soil examination.

Procedure

Plant leaves are carefully examined during a sunny period in midday when transpirational losses are the greatest. Leaves of plants needing water are flaccid, lack turgidity, have a dark, dull appearance, and may be twisted or rolled.

The soil examination is made by removing a sample from the root zone by auger or soil tube. The number of samples examined varies depending upon depth of rooting and soil variation. In taking the soil sample, the presence of compaction or hard layers should be noted. The soil is evaluated by carefully rubbing it between the thumb and forefinger and squeezing a handful very firmly and comparing the findings with values given in Table 7.4.

Gravimetric

Gravimetric procedures work well for many different kinds of soil, but they are time consuming and tedious. Knowledge of the amounts of water held at the wilting point and at field capacity is needed in order to utilize soil moisture as a guide for initiation of irrigation.

Procedure

The total moisture of a soil sample removed from the root zone is determined by weighing out a subsample, allowing it to remain in an oven at 220°F until it loses no more water (24 hours are usually sufficient), and reweighing the sample. Percent moisture is calculated from the equation:

$$(1) \qquad \text{Percent moisture} = \frac{\text{loss in weight}}{\text{oven dry weight}} \times 100$$

This value (expressed as a decimal) is multiplied by the bulk density to determine the volume of water in cubic centimeters per

TABLE 7.4. Available soil moisture as indicated by appearance and behavior when handled

FEEL OR APPEARANCE OF SOIL

Available water*	Sand	Sandy loam	Loam/Silt loam	Clay loam/Clay
Above field capacity	Free water appears when soil is bounced in hand.	Free water is released with kneading.	Free water can be squeezed out.	Puddles; free water forms on surface.
100% (field capacity)	Upon squeezing, no free water appears on soil, but wet outline of ball is left on hand. (1.0)**	Appears very dark. Upon squeezing, no free water appears on soil, but wet outline of ball is left on hand. Makes short ribbon. (1.5)	Appears very dark. Upon squeezing, no free water appears on soil, but wet outline of ball is left on hand. Will ribbon about 1 inch. (2.0)	Appears very dark. Upon squeezing, no free water appears on soil, but wet outline of ball is left on hand. Will ribbon about 2 inches. (2.5)
75-100%	Tends to stick together slightly, sometimes forms a weak ball with pressure. (0.8 to 1.0)	Quite dark. Forms weak ball, breaks easily. Will not stick. (1.2 to 1.5)	Dark color. Forms a ball, is very pliable, slicks readily if high in clay. (1.5 to 2.0)	Dark color. Easily ribbons out between fingers, has slick feeling. (1.9 to 2.5)
50-75%	Appears to be dry, will not form a ball with pressure. (0.5 to 0.8)	Fairly dark. Tends to ball with pressure but seldom holds together. (0.8 to 1.2)	Fairly dark. Forms a ball, somewhat plastic, will sometimes slick slightly with pressure. (1.0 to 1.5)	Fairly dark. Forms a ball, ribbons out between thumb and forefinger. (1.2 to 1.9)
25-50%	Appears to be dry, will not form a ball with pressure. (0.2 to 0.5)	Light colored. Appears to be dry, will not form a ball. (0.4 to 0.8)	Light colored. Somewhat crumbly, but holds together with pressure. (0.5 to 1.0)	Slightly dark. Somewhat pliable, will ball under pressure. (0.6 to 1.2)
0-25% (0% is permanent wilting)	Dry, loose, single-grained, flows through fingers. (0 to 0.2)	Very slight color. Dry, loose flows through fingers. (0 to 0.4)	Slight color. Powdery, dry, sometimes slightly crusted, but easily broken down into powdery condition. (0 to 0.5)	Slight color. Hard, baked cracked, sometimes has loose crumbs on surface. (0 to 0.6)

*Available water is the difference between field capacity and permanent wilting point.
**Numbers in parentheses are available water contents expressed as inches of water per foot of soil depth.
Source: Goldhammer, D. A. and R. L. Snyder. 1989. *Irrigation Scheduling,* Publication 21454. Division of Agriculture and Natural Resources, University of California.

centimeter or inches per cubic inch of soil. The difference between what is held by the soil and its MHC can be applied as irrigation whenever the soil content approaches a critical point.

Tensiometers

A much more rapid method of determining water needs is made possible by the use of tensiometers. These devices consist of a porous cup and a rigid tube filled with water that is attached to a mercury manometer or vacuum gauge. Water can move freely through the porous cup in both directions. As the soil dries, water moves out of the cup, increasing the tension, which can be read from the manometer or the vacuum gauge. The maximum tension that can be read is about 1 bar, or 100 centibars, although bubbles may appear at a tension of 85 centibars, limiting its practical use to only 80 centibars of tension.

Because of an upper practical limit of 1 atmosphere, tensiometers are of limited value for measuring tensions on the heavier loams and clay soils. Because of poor contact between particles and the cup, the use of standard tensiometers is also questionable for measuring tensions in very coarse soils or media containing considerable amounts of coarse particles (anthracite, letite, perlite, coarse sand, and scoria). Successful production with such soils or media is possible if low moisture tensions are maintained by frequent watering, but standard tensiometers are not suitable in detecting small differences in the 8-15 centibar range at which point irrigation may be needed. A new type of "Low Tension" tensiometer (Irrometer model LT) with an effective 0 to 40 centibar range, overcomes this difficulty, making it practical to monitor moisture in coarse soils or nonsoil media mixes.

Tensiometer Placement

Most accurate measurements are made within 2 inches of the cup, making placement in relationship to root distribution an important factor. The use of tensiometers is highly desirable for nursery, greenhouse, agronomic, and vegetable crops with their limited root systems. It is less reliable for orchard crops, where accurate placement of the cup in relationship to the most active roots becomes more of a problem.

The depth of cup placement needs to vary with crop, soil, and method of cultivation. Instruments are available in standard lengths of 6, 12, 18, 24, 36, 48, and 60 inches. Use of the 6-inch plus 18-inch lengths are adequate for shallow-rooted crops, but 12-inch plus 24-inch lengths need to be used for deeper-rooted crops. Another placement at 36, 48, or even 60 inches may be desirable for very deep-rooted crops. Placement must always consider root distribution as well as depth of penetration. Recommended depths of placement are given in Table 7.5.

A correction is needed to compensate for the gravitational potential of deep-placed tensiometers. The correction can be made on some gauges by turning the calibration screw, but for others it will be necessary to subtract three centibars for each foot of depth.

The placement must ensure continuous contact between the soil and the porous cup. The diameter of the hole punched or augered for placement of the tensiometer should be equal to the diameter of the cup. The soil around the cup needs to be moistened to make good contact with the cup.

Several placements must be made in each field, increasing the number as the variability of the soil increases. Representative sites are selected, avoiding unusually low, wet areas or high, dry areas. Placement in areas of variable soil conditions is of the utmost importance when tensiometers are used to activate irrigation automati-

TABLE 7.5. Recommended placement depths for tensiometers

Depth of root system	Shallow instrument	Deep instrument
in.	in.	in.
Up to 18	8-12	—
Up to 24	10-14	24
Up to 36	14-18	36
Up to 48	18-24	36-48
Over 48	24	48-60

Source: Irrometer Co., Riverside, CA.

cally. In such cases, the irrigation system needs to be engineered to irrigate different sections separately.

Tensiometers can be broken and must be placed in the field so that they will not be disturbed by equipment or people. Placement within a row is usually satisfactory for annual crops that are irrigated by furrow or overhead sprinkler. For the sprinkler irrigation systems, the tensiometers should not be placed directly under a sprinkler head or permanent lines where the soil is usually moister. Nor should they be placed where they may be shielded from the water by plant cover. In recently set orchards, the porous cup is placed in the root ball, but it is moved to other locations and others added as the tree grows. New locations are chosen to reflect the changes in root distribution, but placement in areas that are heavily shaded or receive unusual amounts of water–either too little due to limb blockage or an excess from the drip line–must be avoided. For most situations in northern latitudes, ideal placement will be on the southwest side of the tree near (but not at) the drip line. Because of differential wetting patterns produced by drip irrigation in different soils, the tensiometer cup is placed about 12 inches from emitters on sandy soils and about 18 inches away for the heavier soils.

Preparation and Maintenance of Tensiometers

Tensiometers need to be prepared and checked prior to installation. They should be filled with air-free water (boiled water) or the air must be removed by a hand vacuum pump. The instrument is checked by allowing the porous cup to gradually dry in the air so that a high tension develops. The tension should drop to zero within a few minutes after placing the cup in water.

Proper maintenance is required for the instrument to function properly. Because air is extracted from water under tension and can be trapped in the tensiometer, it is necessary to periodically remove the cap and refill the tube with water. The need for refilling with water, signaled when more than 0.5 inches of air has accumulated under the cap, is greater with dry soils than with wet soils. Large bubbles, arising as suction is applied by a hand pump, indicate the presence of leaks which will have to be corrected. Usually such leaks occur at the bottom of the tensiometer, but they may also be in the gauge or around the stopper. Occasionally, the ceramic tips will have to be replaced as they become clogged with calcium or saline deposits. Tensiometers should be removed or alcohol added to the water in the porous cups under conditions where freezing is a possibility.

Readings

Instrument readings are taken daily, preferably at the same time each day, and a record is kept of the daily readings at various locations. Plotting the daily readings will enable rapid visualization of changes taking place and alert the grower to the need for irrigation.

Saturation, or free water, is indicated by a zero reading. Field capacity or MHC are indicated by readings of 10 centibars for sands to 30 centibars for clay soils. Unless there is a need to leach out salts, avoid irrigating beyond field capacity in order to conserve water and eliminate leaching losses, which can pollute water supplies.

The readings at which irrigation should start vary with different crops and their stages of growth. (See Table 7.6.)

Irrigation should be started at low soil moisture tension (20-30 centibars) for young or shallow-rooted crops on soils of low MHC, or for critical periods of growth. Frequent small applications are more suitable for shallow-rooted crops grown on light soils than heavy waterings applied less frequently. Greater tensions (40-50 centibars) can be tolerated for crops with deeper roots, especially if they are grown on the heavier soils.

Starting irrigation at lower soil moisture tensions is also advisable under conditions of high evaporation (high temperatures with ample solar radiation and/or strong winds). The upper cups (6-12 inches) can be used to start the irrigation, but the readings of the lower cups are used to help evaluate the desirable length of the irrigation period.

The scheduling of the irrigation is often dictated by the capacity of the system. If all the areas cannot be watered at the same time, irrigation may have to be started earlier on a portion of the farm in order to insure that all of it is irrigated before harm is done to some of the crop.

Greater accuracy in timing the start and duration of the irrigation is possible if a curve is drawn for each field showing the moisture contents at different tension readings. Samples of soil taken at cup placement depth are collected over a period of time under con-

TABLE 7.6. Soil tensions* at which several crops need to be irrigated

Crop	Tension	Crop	Tension
	cbrs.		cbrs.
Avocado	50	Grape	40-80
Beet, sugar	60-80**	Lemon	40
Beet, table	40-60	Lettuce	40-60
Broccoli, early	45-55	Onion, dry	55-65
Broccoli, late	60-70	Onion, green	45-65
Carrot	55-65	Orange	20-80
Cabbage	60-70	Potato, Irish	30-50
Cantaloupe	35-40**	Potato, sweet	40-70
Cauliflower	60-70	Rhubarb	60-85
Celery	20-30	Sorghum	60-85
Deciduous fruit	50-80**	Watermelon	50-85
Endive	30-40		

* Where two readings are given, the lower one should be used for conditions of high evaporation, and the higher value for low evaporative conditions.
**Ideal centibar readings to start irrigation during the period when the crop approaches maturity are about 80 for sugar beets, 75 for cantaloupes, and 80 for deciduous fruits.

Source: Original data presented by Goldberg, D., B. Gormat, and D. Rimon. 1976. *Drip Irrigation*. Drip Irrigation Scientific Publications, Kfar Schmaryanu, Israel. Modified by the author.

ditions varying from full capacity to readings of about 80 centibars. Moisture in the samples is determined gravimetrically and converted to percent volume. These values are plotted against centibar readings existing at the time of sample collection. (See Figure 7.1.)

Gypsum Blocks

Soil moisture can be evaluated by measuring the resistance between two electrodes planted in gypsum. Because of rapid deterioration of the gypsum blocks in the soil, the all cast-gypsum blocks have given way to gypsum combined with nylon or fiberglass. More recently, a casting with a nylon unit inside the gypsum blocks has been used. Water moves in and out of these blocks very much as it moves in the soil.

Figure 7.1. Relationship of soil moisture contents* of a sandy soil to different tensiometer readings taken at a depth of 6 inches

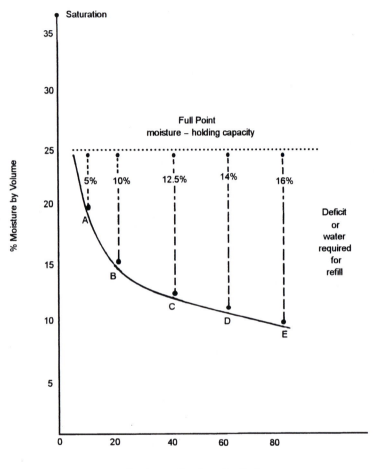

* Percent by volume

Gypsum blocks are more accurate than tensiometers in ranges of high tensions (greater than 0.33 bars) and, therefore, are better-suited for medium- and heavy-textured soils. Nevertheless, they are considered less reliable than tensiometers under conditions that include: (1) sands or coarse textured soils that release considerable water at low tensions; (2) drip irrigation, due to their poor sensitivity at high moisture levels; and (3) saline-alkaline soils, as salts affect resistance. Use of screen or cylindrical electrodes reduces the salinity effect and can make gypsum block use in saline-alkaline soils more accurate.

Preparation and Placement

Blocks are tested by soaking in water and hooking them up to a resistance meter before installation. A functioning, single block is placed in a hole only slightly larger than the block. Making the hole about 1 inch deeper than the intended placement allows for the addition of loose soil and a little gypsum with lime under the block. The block is placed firmly in the hole, and the soil mixture with gypsum is added along the sides as necessary. Adding several fluid ounces of water around the block helps to seal it in place. The block is covered with the soil, gypsum, and lime mixture for several inches and the remainder of the hole is filled with soil, tamping carefully to avoid damage to the block or wires. Blocks, like tensiometers, are placed at different locations in the field and at different depths depending upon the rooting system and type of soil.

Readings

Wires from the blocks are brought to the soil surface. Readings are taken with a portable resistance unit (soil moisture meter) that is specially calibrated and can be used to service many blocks. Meters can be calibrated to read directly in percentages of available soil moisture. Otherwise, the resistance in ohms can be converted to percent moisture by plotting the measurements of resistance as ohms, with amounts of readily available moisture varying from field capacity to the wilting point. A similar type of curve is drawn of any new soil even if the instrument readings are given as percentages. (See Figure 7.2.)

FIGURE 7.2. Percentages of soil moistures corresponding to resistance in ohms of several soils and their need for irrigation

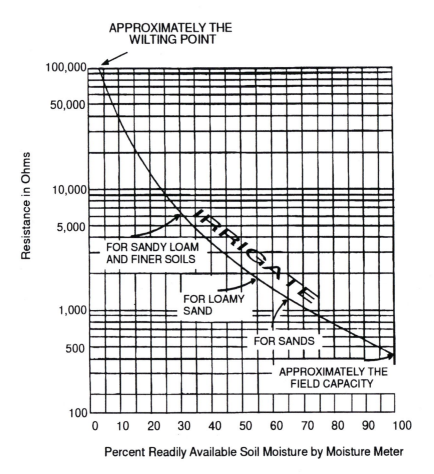

Percent Readily Available Soil Moisture by Moisture Meter

Source: Bouyoucos, G. J. 1956. Improved soil moisture meter. *Agricultural Engineering* 37(4): 261-262. Reprinted by permission of American Society of Agricultural Engineering.

Thermal Change Sensors

The conduction of heat by soil is reduced as moisture content decreases. Initially, this principle was applied to soil moisture evaluation by winding enamelled copper wire around pieces of glass tubing and burying the unit in the soil. A controlled current was passed through the wire for one to two minutes. The resulting rise in temperature was measured by changes in the wires' conductivity.

Newer units, using the same principles have been enclosed in plaster of Paris blocks to reduce variability resulting from poor contact between soil and heating unit, but these blocks were less satisfactory than ordinary gypsum blocks in measuring high water tension. Further improvements utilize porous ceramic cups or disks to enclose the electrical unit. Water flows in and out of the cup or disk, maintaining equilibrium with water in the soil. The pore size of the disks can be varied for soils of different textures. Due to variation in pore size and electrical components, each unit needs to be standardized with a pressure plate apparatus or with other known standards.

The soil moisture is monitored with the following equipment: (1) a battery for power supply; (2) a precise 10 K resistor; (3) a 4 1/2-digit voltmeter capable of measuring microvolts in the 0-19 millivolt range; and (4) a timer. Several commercial units comprising these various parts are available at costs of under $2000.

Thermal units lend themselves to automatic irrigation control. For automatic use, additional electronic capability is needed to: (1) automatically sample several sensors sequentially; (2) compare the sensors' output to threshold values for starting irrigation, and (3) to start the pump and open valves as needed.

Procedure

A controlled current is passed through the wire for one to two minutes, and the resulting rise in temperature is measured by changes in the wires' conductivity. The readings are compared to those of soils with known moisture contents. A curve showing readings taken at field capacity and at various stages of drying until wilting occurs, with moisture contents as measured by gravimetric means, serves as an excellent standard.

Neutron Probe

The energy loss of fast neutrons from radioactive cesium or americum as they collide with slow neutrons from water forms a basis for measuring available water. A suitable instrument for measuring moisture will consist of a fast neutron source, a slow neutron detector, and a counting device. A scaler used with the soil probe is unaffected by the fast neutrons, and reacts only with slow neutrons. The fast neutrons collide with the hydrogen of water and are gradually converted to slow-moving neutrons that can be counted by the detector. The number of slow-moving neutrons counted in a given time period indicates the amount of water present, because hydrogen in water molecules is the primary atom of low atomic weight in soils.

Readings need to be correlated with actual moisture contents of different soil types, collected at various depths and known volumes. The simplest way of doing this is to collect samples of soil at various depths as the neutron access tube is placed. The moisture content of these samples is then correlated with neutron probe readings taken at the same depths. Evaluation of available moisture is very rapid once the calibration curves are established from these readings. The primary disadvantages of this method are: (1) the cost, which can be several thousand dollars; and (2) the expertise needed to handle sensitive radioactive equipment.

Placement and Readings

Readings are taken by lowering the probe to the desired depth in previously placed access tubes. Tubes made from aluminum are preferred, although steel, thin iron, and plastics have been used. A series of tubes (2-inch outside diameter) are installed vertically in holes bored by an auger of the same diameter to a depth 12-inches deeper than expected roots. The holes are located at strategic sites representative of the field to be irrigated, but should not be within 16 feet of any other neutron source. The tubes are tapped lightly and left in place, extending 2-10 inches above the ground. Readings are taken after approximately 30 seconds with a unit that can effectively service many locations. Readings are compared to standards set up at the time of installation.

Capacitance

A nonnuclear soil probe using an access tube employs capacitance technology to measure the soil moisture by measuring changes in the dielectric constant of the soil. Most solid materials in the soil (sand, silt, clay, and OM) have dielectric constants ranging from 2 to 4, while that of water is 78. A 2-inch PVC access tube placed in the soil is used and the probe is lowered in the access tube. Readings are taken at varying depths to determine the moisture profile. Readings can be compared with factory preset calibrations, or more accurately with values calibrated for the particular soil. These can be obtained by measuring soil moisture levels of samples collected at similar depths and determined by pressure plate or gravimetric means.

The method is very fast and works well over a wide range of temperatures (32°-158°F) and moistures (0-60 percent volume). The Sentry 200-AP, manufactured by Troxler Electronic Labs, Inc. Research Triangle Park, NC 27709, uses a probe weighing a little over 2 pounds and a control unit that weighs about 8 pounds. Costs for both units are about $3700.

Evapotranspiration

Most soil water is lost by evaporation from the soil surface or by transpiration from the leaves. The two sources of loss, which are difficult to measure separately, are considered as a unit designated as evapotranspiration, or Et. Water loss through Et is influenced by heat, day length, air humidity, radiation, type of plant, and the plant's coverage of the soil surface. Using Et data for irrigation control is reliable and practical, but requires careful monitoring. The costs of the pan and its installation can be depreciated over a number of years.

Pan Evaporation and Et

The amount of water lost by evapotranspiration is directly proportional to that evaporated from a standard pan. The amount lost from a standard pan will be equal to evaporation if: (1) the soil is wet;

(2) plants cover the soil completely; and (3) the pan and plants are exposed to exactly the same conditions of wind and sunlight. Since all of these conditions are seldom met, a correction factor is needed to convert pan evaporation (Ep) data to evapotranspiration (Et).

In the U.S.A., a standard pan (Class A) is a circular pan, made of unpainted galvanized iron, with an inside diameter of 120 centimeters (48 inches) and 25 centimeters (10 inches) high. The unit is supported 15 centimeters (6 inches) above the ground by wooden planks to allow for free air movement.

The water must be kept clean. A wire screen can be used to protect the pan from birds and animals, but Ep values are lower by about 10.4 percent as compared to those for an unprotected pan.

The depth of the water is maintained at 18-20 centimeters (4-8 inches), with fluctuations no greater than 5 centimeters (2 inches). The height of the water is accurately measured on a daily basis at the same time of day with a micrometer depth gauge accurate to 0.01 millimeters or 0.1 inches. This reading will indicate the 24-hour net change in water as affected by rainfall and evaporation.

This Ep data can be converted to Et by use of the equation:

(2) $Et = Ep \times F \times C$
            Where F = crop factor
            and C = leaf coverage (solid cover = 1.00)

Correction Factors (Et/Ep)

The correction factors for converting the Ep of pans placed in the open to Et varies from a factor of about 0.85 for conditions of high humidity and low wind to about 0.50 for conditions of low humidity and high wind. An average value of 0.7 can be used as a starting point if there is good penetration of water into the soil.

Correction factors have been calculated for different stages of growth of a number of crops in several localities and often this data can be obtained from the horticultural or engineering departments of the state land grant colleges or from Extension Services within the state.

The assumed factors (0.60 or 0.70 for good infiltration, or 1.0 for ponded soils or those determined for particular crops by state land grant colleges) can be refined to represent specific conditions by combining pan data with data from tensiometers. The daily readings of tensiometers, placed as suggested earlier, are used to evaluate whether the assumed crop factor is maintaining satisfactory moisture. If moisture cannot be maintained in the 10-20 centibar range by irrigation based on the assumed factor, the factor is changed upward if the tensiometer readings increase and downward for lowered tensiometer readings. The assumed factor is correct if tensiometer readings are maintained in the desired range.

Checkbook Method

Introduction

The net loss of water (rainfall − Et) is subtracted from amounts held by the soil (MHC) to determine when irrigation should start and what amount should be applied. The usefulness of such calculations is enhanced if data providing effective monthly rainfall is included. The latter is calculated from a formula that incorporates total monthly rainfall, monthly consumptive use, and the soil storage factor. A less accurate method, using generalized Et data for particular crops in an area combined with specific rainfall and moisture holding data, can also be utilized to provide the "checkbook" entries.

Computer-Controlled Data

Variations of this concept are used in computer programs. Various formulas have been developed over the years predicting Et from climatic data, such as mean monthly temperature, daytime hours, and humidity. Consumptive data for different crops have been calculated from monthly temperature, percentage of daylight hours, precipitation, frost-free growing period, and irrigation data. The seasonal requirement depends on the consumptive use minus rainfall or irrigation multiplied by a farm irrigation efficiency factor.

Reference evapotranspiration (Eto) and crop co-efficients (Etc) for different crop stages has been calculated for some regions and this data often may be estimated for other areas by interpolation from isoline maps provided by some state Extension Services.

In California, a computer program supplies such data through the California Irrigation Management Information System (CIMIS), with direct access to the grower by telephone. Daily weather and water supply (rainfall and irrigation) data helps provide a refined irrigation schedule. The service is available through the California Department of Water Resources, Office of Water Conservation, P.O. Box 942836, Sacramento, CA 94236-0001.

A somewhat similar service is offered by the University of Wisconsin Extension Service and is designated as WISP. The water retained by the soil and available to crops, known as allowable depletion, has been calculated for different soil types and crops. The computer calculates the amount and percentage of allowable depletion from farmers' data for irrigation and rainfall, and estimates the allowable depletion in 24 or 48 hours if the crop receives no irrigation or rain.

The need for irrigation as indicated by CIMIS, WISP, and related services may have to be modified to allow for infiltration rate, runoff, irrigation uniformity, and potential contributions from water tables, dew, and precipitation, as these factors influence irrigation efficiency and the need for water.

Difference Between Foliar and Air Temperatures

Regulating irrigation by the use of foliar temperature differentials is based on the concept that leaf temperatures begin to rise as the moisture necessary for maintaining the transpiration stream is limited.

The method offers great promise for scheduling irrigation starts in arid or semi-arid climates, but appears to have less value in a high-humidity climate or where rain falls during the growing season. But even in the more arid climates, practical use for many crops awaits the determination of satisfactory cumulative thresholds.

Procedure

Leaf and air temperatures are measured by a small portable infrared thermometer on a daily basis. Measurements are commonly made at one hour past solar noon with the back to the sun and the instrument held at a 45° angle to the leaf canopy. Positive values (leaf temperatures greater than air temperatures) are recorded and accumulated. The need for irrigation is signaled when the accumulated stress degrees (degrees of canopy temperature which are greater than the ambient temperature) exceed the threshold value. At just what point increases in leaf temperatures cause enough stress to warrant irrigation is still being determined for many crops at various stages of growth. The threshold value for wheat is about 15 accumulated stress degrees.

ADDING THE CORRECT AMOUNT OF WATER

The amount of water that has to be added to promote ideal growth will vary with the crop, the stage of the crop, the soil, present water content, and the depth of rooting.

Calculating Amounts Needed

An approximation of the amounts needed for different soil textural classes to reach field capacity is presented in Tables 7.1 and 7.2 (pp. 169 and 170-171).

A more accurate estimation can be obtained by calculating the amounts of available water and subtracting this amount from the amount of water held at field capacity. Available water is calculated as follows: (1) Bulk density of the soil is determined (see Chapter 2); (2) the percent water (weight basis) held at field capacity and at the wilting point is converted to volumes of water by multiplying the percent of water by the bulk density; and (3) the amounts of water being held by a centimeter or an inch of soil are converted to the amount held by the effective root zone. For example, if the percent of water at field capacity equals 25.6, that at the wilting point equals 10.5, and the bulk density equals 1.27 g/cc, then the volume percent of the available water equals 0.256 minus 0.015 multiplied by 1.27, or 0.192 centimeters of water per centimeter of soil. This is equivalent to 0.192 inches of water per inch of soil. The available storage capacity of an effective root zone can be obtained by multiplying a weighted average capacity by the root depth, or adding together the amounts of water for all soil depth increments.

If tensiometers are used, the amount of water needed at various tensions can be read as the difference between soil moisture present and that needed to bring it to full capacity. For example, in Figure 7.1, showing typical moisture percentages at different tensiometer readings, 10 percent of the water volume is deficient at point B. If the cup was placed at a 6-inch depth, it would be necessary to add 0.60 inches of water (0.10 × 6). Since there are 27,154 gallons of water in an acre inch, the 0.6 inches of water would require 16,292 gallons per acre.

Water Required and Et/Ep

The amount of water needed can be calculated from Et/Ep relationships and tensiometer readings. If an assumed 70 percent factor is correct (relationship of Et to Ep), each inch (2.5 centimeters) of Ep will require 0.7 inches of applied water to keep the tensiometer readings in the 10-20 centibar range. The 0.7 inches of water need to be applied overall only if there is solid cover. For irrigation systems that can supply the water to limited areas, the percentage leaf cover will have to be calculated for less-than-solid cover. The percentage leaf coverage for orchard crops is the ratio of canopy area to total area of the tree spacings. For example, if tree canopy is 10 × 14 feet and the trees were spaced 15 × 18 feet, the leaf coverage (C) equals:

$$(3) \qquad C = \frac{10 \times 14}{15 \times 18} = 0.518, \text{ or } 51.8 \text{ percent}$$

For row crops, the coverage is computed from the ratio of the width of the leafy area to row spacing. For plants having a width of 12 inches and a row spacing of 36 inches, the ratio is:

$$(4) \qquad C = \frac{12}{36} = 0.333, \text{ or } 33.3 \text{ percent}$$

Using equation #1 in this chapter, the water needed for trees with canopies of 10 × 14 feet in spacings of 15 × 18 feet when the Ep equals 0.4 inches per day and the factor equals 0.75 will be:

(5)   Et = 0.4 × 0.75 × 0.518, or 0.155 inches per day

The water needed for row crops with a canopy 10 inches wide in rows 36 inches apart, with an Et of 0.4 and a factor of 0.75, can be calculated from equation #2. Entering the several components into the equation gives:

Et = 0.1 inch per day derived from (0.4 × 0.75 × 0.33)

The Et can be converted to gallons per tree (GPT) for complete coverage by using the following equation:

(6)   GPT = 0.623 × Et × plant area (tree setting)

Using the data given for #3, the gallons needed daily per tree are:

GPT= 26.07 derived from (0.63 × 0.155) × (15 × 18)

The gallons per acre (GPA) for solid coverage can be calculated by multiplying the Et in inches per day by 27,154 (the gallons per acre inch). An Et of 0.145 will require 3937 gallons per acre each day. For less than solid coverage, the 27,154 must be multiplied by the percentage cover (from equations #3 or #4).

Chapter 8

Water Quality

MAINTAINING WATER QUALITY

Poor water quality can adversely affect the operation of some water systems, harm crop production, and in time lead to soil degradation.

Quality Tests

The suitability of irrigation water for various purposes can be ascertained by a series of chemical tests. (See Table 8.1.) Because water quality must be better with small orifice systems (drip irrigation), amounts of suspended solids and number of bacteria should be determined for these systems. (See Table 8.2.)

Suspended Solids

Water with large amounts of suspended solids must be held in a settling pond and passed through screens to reduce the quantities. Water intended for the micro-irrigation systems must have even fewer suspended solids, and so will have to be passed through filters in addition to screens.

Microorganisms

The presence of large numbers of bacteria is troublesome, primarily for micro-irrigation systems where they create various slimes that block emitters. Chlorine is routinely added to waters used for systems with micro-orifices to destroy bacteria and other microorganisms, reducing the practical need for bacterial counts.

The efficiency of the chlorination can be measured by placing a removable glass slide in a PVC fitting in the water flow path, remov-

TABLE 8.1. Irrigation water quality based on chemical tests

Water quality parameter			Severity of potential problems		
			None	Increasing	Severe
Acidity					
pH	(a)		5.5-7.0	<5.5 or >7.0	<4.5 or >8.0
Salinity					
Conductivity EC mmhos/cm			0.5 to 0.75	0.75-3.0 (b)	>3.0 (b)
Dissolved salts		ppm	320 to 480	480-1920 (b)	>1920 (b)
Permeability					
Caused by low salts					
Conductivity EC mmhos/cm			>0.5	0.5-0.2	<0.2
Dissolved salts		ppm	>320	320-125	<125
Caused by excess sodium					
SAR	(c)		<3.0	3.0-6.0	<6.0
Toxicity					
Bicarbonate	(d)	ppm	<40	40-180	>180
Boron		ppm	<0.5	0.5-2.0	>2.0
Chloride	(e)	ppm	<70	70-300	>300
Fluoride	(f)	ppm	<0.25	0.25-1.0	>1.0
Sodium	(g)	ppm	<70	70-180 (b)	>180 (b)
Clogging	(h)				
Calcium	(i)	ppm	20-100	100-200	>200
Carbonates		ppm	<35	35-75	>75
Magnesium		ppm	<63	>63	
Iron	(j)	ppm	<0.1	0.1-0.4	>0.4
Manganese		ppm	<0.2	0.2-0.4	>0.4
Sulfides		ppm	<0.1	0.1-0.2	>0.2
Nutritional					
Nitrogen	(k)	ppm	<5	5-30	>30

(a) The water pH affects pH of low-buffered media. High pH aggravates emitter clogging and foliar staining. Low pH increases corrosion of metal pipes and equipment.

(b) The conductivity and amount of dissolved salt causing problems are affected by soil class and crop tolerance to salts.

(c) Sodium absorption ratio. (Discussed in this chapter.)

(d) Bicarbonates cause problems by modifying soil calcium. If applied overhead, phytotoxicity can also be a problem.

(e) Amounts of chlorides that are phytotoxic to some plants when applied overhead may not be harmful if soil applied.

(f) These values are significant for fluoride-sensitive plants.

(g) Less severe if potassium is present in equal quantities or if plants are tolerant to sodium.

(h) Clogging is more serious in system with small orifices.

(i) Calcium causes clogging as phosphates or sulfates are added in irrigation lines, particularly if the pH is high.

(j) Values >0.2 ppm increase clogging from bacterial slimes and can cause staining of plants as water is applied overhead. Concentrations of 0.4 ppm can cause sludges if chlorine is used.

(k) Values >5 ppm can stimulate algae growth in ponds. Values >30 ppm can delay maturity of some crops and decrease sugar content of crops such as sugar beet. For nonsensitive plants the N content can be beneficial but should be considered in the fertilization program.

TABLE 8.2. A system for classifying water for drip irrigation systems

Rating	Suspended solids	Dissolved solids	Bacteria
	ppm	ppm	no/ml
0 – ideal	<10	<100	<100
1	20	200	1000
2	30	300	2000
3	40	400	3000
4	50	500	4000
5	60	600	5000
6	80	800	10000
7	100	1000	20000
8	120	1200	30000
9	140	1400	40000
10 – unsatisfactory	>160	>1600	50000

Source: Bucks, D. A., E. S. Nakayama, and R. G. Gilbert. 1979. Trickle irrigation water quality and preventative maintenance. *Agricultural Water Management* (2):149-162. Reprinted by permission.

ing it periodically to examine it for gels or slimes, which indicate an undesirable number of bacteria.

The presence of sufficient chlorine to keep microorganisms under control can be determined by chlorine titrators, or more rapidly by various kits. Adequate control is obtained with 1-2 ppm "free residual" chlorine at the farthest outlet from the injection point. Kits based on the N, N-Diethyl-p-Phenylenediamine (DPD) procedure are satisfactory, but ordinary swimming pool kits using ortho-tolidime to measure total chlorine are not.

Chlorine

The following chlorine concentrations (measured at the end of the farthest lateral) and application timings are recommended for different problems.

1. Continuous injection to supply 1-2 ppm to control bacteria and algae.
2. Intermittent injection of 10-20 ppm for 30-60 minutes to kill a buildup of algae or bacteria.
3. A one-time application of 500 ppm to clear blockages caused by organic matter. The system is shut down for 24 hours, allowing the 500 ppm chlorine to destroy the organic matter, after which time all submains and laterals are thoroughly flushed.

Several chlorine sources can be used. Chlorine in the form of a liquefied gas under high pressure is the most economical, but due to its highly poisonous and corrosive qualities is seldom used except for large operations. Solid calcium hypochlorite or liquid sodium hypochlorite formulations are more suitable for smaller systems. The chlorine equivalents of the different materials are given in Table 8.3.

Calculations for injecting the various forms of chlorine are as follows*:

 a. Gaseous chlorine Cl_2. A chlorinator is used to meter gas into the supply line. Its injection rate is calculated from the equation:

(1)   IR $= Q \times C \times 0.012$
 Where IR = chlorine injection rate (lb/day)
      Q   = system flow rate (gpm)
      C   = desired chlorine concentration (ppm)

 b. Liquid sodium hypochlorite (NaOCl) is metered into the system, calculating the injection rate from the equation:

(2)   IR $= Q \times C \times 0.006/S$
 Where IR = chlorine injection rate (gals/hr)
      Q   = system flow rate (gpm)
      C   = desired chlorine concentration (ppm)
      S   = strength of NaOCl solution (%)

 c. Calcium hypochlorite (Ca[OCl]$_2$) is dissolved in water before injecting into the system. Each percent chlorine

---

*From Boswell, M. J. 1990. *Micro-Irrigation Design Manual*. Fourth Edition. James Hardie Irrigation, Inc., El Cajon, CA 92020.

solution made from the 65 percent material requires 12.8 pounds of calcium hypochlorite per 100 gallons of water, derived by dividing the weight of a gallon of water (8.34) by the percentage of chlorine (as a decimal) in the dry calcium hypochlorite.

The injection rate is determined by figuring the percentage of the prepared calcium hypochlorite solution into equation #2 used to calculate the rate for sodium hypochlorite.

Algae in Surface Waters

Algae in ponds, reservoirs, and lakes that can cause problems with filters can be controlled by the addition of 1-2 ppm of copper sulfate (bluestone). Applications are limited to water with temperatures of at least 60°F. Because growth of algae is dependent on light, only the upper 6 feet of water needs to be treated. The copper sulfate can be added by: (1) placing it in open mesh bags suspended

TABLE 8.3. Sources of chlorine useful for microbiological control of irrigation water

Chlorine source	Equivalent amounts	Amount per acre foot*
Chlorine gas 100% Cl_2.*	1.0 lb	2.7 lb
Calcium hypochlorite 60-70 Cl_2**	1.5 lb	4.0 lb
Sodium hypochlorite		
15% Cl_2**	0.8 gal	2.2 gal
10% C_{l2}**	1.2 gal	3.3 gal
5% Cl_2**	2.4 gal	6.5 gal

* To obtain 1 ppm of chlorine at injection point.
** Available chlorine.

Source: Boswell, M. J. 1990. *Micro-Irrigation Design Manual.* Fourth Edition. James Hardie Irrigation, Inc., El Cajon, CA 92020. Reprinted by permission.

in water and kept in place by tying to stakes; (2) dragging sacks of crystals suspended in water behind a boat traveling parallel paths about 15 feet apart; or (3) sprinkling 16-mesh crystals into the wake of a boat as it makes the parallel passes.

Copper sulfate can be toxic to fish. Never treat more than 1/2 of the area at one time so that fish can move to untreated areas.

Amounts of copper sulfate that can cause toxicity to fish are dependent upon water acidity. The 2.7 pounds of copper sulfate per acre foot of water required to obtain 1 ppm when alkalinity value (calcium carbonate in mg/l) is over 100 needs to be reduced to 1.3 pounds if alkalinity value is 60-100 and to only 1 pound per acre foot if the alkalinity value is 40-60. To avoid killing fish, copper sulfate should not be used if the alkalinity value is less than 40. Most laboratories can provide alkalinity values by titrating the water with sulfuric acid.

CHEMICAL EVALUATION

Collection of Samples

Chemical evaluation is made on two samples from the same source, each consisting of about 1 quart of water. Samples are placed in clean plastic or glass containers that have been rinsed several times in the test water prior to filling. Water from a domestic tap line or well is allowed to run for at least 15 minutes before collecting samples. Samples from open bodies of water are collected beneath the surface and near the point of water intake. About ten drops of hydrochloric (muriatic) acid are added to one of the filled containers for iron (Fe) and manganese (Mn) analysis. The water of the other container is used for all other tests.

Procedures

Methods of analysis are quite similar to that for soils, except no extraction is necessary. Conductivity and pH meters are used to test for salts and pH. Generally, a simple filtration prepares the sample for either colorimetric, atomic absorption, or plasma analysis. Adding a small amount of activated charcoal may be necessary to decolorize some water samples before preparing them for colorimetric analysis.

The importance of various characteristics of water, and some means of modifying adverse conditions, are covered in some detail in the sections below.

pH

A pH range of 6.0 to 7.0 is considered ideal because this range is satisfactory for growing a large number of crops. Lower values (to about 5.0) are preferred for acid-loving plants, but the lower values can cause corrosion of several metals used in water-moving equipment. Very low or very high water pH values can affect the pH of poorly buffered media (sands, artificial media, and water) enough to have an impact on the availability of several nutrients.

Short-term use of acid waters has negligible effects on buffered soils (good to high CEC), but in time can lower the pH enough to increase aluminum to a level that is toxic to sensitive plants. Some plants that are sensitive to aluminum are listed in Table 8.4.

The pH level affects the solubility of various compounds, whether they are found naturally in the water, added as fertilizer salts in irrigation water, or formed as fertilizer salts react with each other or with the water components. High pH values favor, and low pH values tend to prevent, the formation of many precipitates. If precipitates are allowed to form, they can affect the availability of added nutrients and can reduce the flow of irrigation by clogging small orifices or building up in pipes.

If resistant metals or plastics are used to move water, pH values down to about 4.5 can be beneficial for the introduction of fertilizers which are capable of producing precipitates with water components or with other fertilizer salts (phosphates, sulfates, and calcium nitrate). Acid waters are also useful for cleaning lines and emitters, destroying bicarbonates, and keeping manganese oxides soluble.

Changing Water pH

High pH values can be lowered by the injection of acids. Sulfuric acid is commonly used for pH reduction because of costs, although

TABLE 8.4. Plants sensitive to relatively low levels of several elements

Aluminum	Boron	Chloride	Fluoride	Sodium
Alfalfa	Apple	Avocado	Asparagus,	Almond**
Asparagus	Apricot	Apple	Sprengeri	Apple
Barley*	Artichoke,	Bean,	Apricot,	Apricot**
Beets,	Jerusalem	Broad	Chinese	Avocado
Sugar, Table	Avocado	Dwarf	Blueberry	Banana
Blackberry	Bean,	Navy	Box Elder	Bean,
Bluegrass,	Cowpea	Runner	Coffee,	Field
Kentucky	Elm	Currant	Arabica	Kidney
Cabbage	Kidney	Endive	Cordyline,	Lima
Carrot	Navy	Flax	Baby Doll	Snap
Cantaloupe	Snap	Grape	Ti plant	Celery
Cauliflower	Blackberry,	Lettuce	Corn, Sweet	Cherry, Citrus
Celery	Thornless	Mulberry	Crocus	Clover,
Clover,	Cherry	Onion	Dracaena,	Alsike
Ladino	Chrysanthemum	Pea	Janet Craig	Crimson
Red	Fig, Kadota	Peach	Striped	Dutch
Sweet	Grape	Pear	Warneck	Ladino
White	Grapefruit**	Potato	Elm, English	Red
Lettuce	Larkspur	Red Clover	Fir, Douglas	White
Onion	Lemon**	Strawberry	Freesia	Cucumber
Parsnip	Pansy		Gladiolus	Pea, English
Pea, English	Peach		Grape	Pear
Soybean*	Pear		Jerusalem Cherry	Plum**
Spinach	Pecan		Larch, Western	Raspberry
Strawberry	Persimmon		Lilies,	Strawberry
Timothy	Plum		Ace	
Wheat*	Soybean		Croft	
	Strawberry		Easter	
	Tobacco		Hybrid	

Aluminum	Boron	Chloride	Fluoride	Sodium
	Violet		Peach	
	Walnut, English		Pines,	
	Zinnia		Eastern white	
			Loblolly	
			Lodgepole	
			Mugo	
			Ponderosa	
			Scotch	
			Plum, Bradshaw	
			Prune, Italian	
			Sorghum	
			Spider plant	

* Some varieties
** Greater injury from leaf applied irrigation water

phosphoric, nitric, or citric acid can also be used. If phosphoric or nitric acid is used, the impact on nutrient supply must be considered.

The acid is usually injected throughout the entire irrigation period, although it can be limited to a short period toward the end of the irrigation cycle if clogging of emitters by bicarbonates is the only concern. Acid injection requires that all contacted equipment be acid resistant. Personnel must be knowledgeable in the handling of acid and be equipped with safety clothing and goggles.

The addition of acid into irrigation lines will cause fewer problems if it is diluted to about 10 percent strength prior to injection. Dilution should always be carried out by adding acid to water, and not vice versa, in order to minimize splashing. Using a piece of resistant pipe (about 20 feet in length) just beyond the injection point allows for sufficient mixing of acid and water to minimize corrosion of irrigation pipe, gate valves, and emitters downstream. Resistant metal or plastics need to be used for the entire system if the anticipated pH is less than 6.0.

The amount of acid required to change the water pH to the desired level is most accurately determined by titrating the water with acids. This can be done in most laboratories. A rough estimate can be obtained by filling a 50-gallon drum with water and adding increasing amounts of a dilute acid. The desired endpoint can be determined by periodically measuring the pH of small samples of water with a pH test kit or meter. A suitable dilute acid of 5 percent is obtained by adding 1 fluid ounce of acid to 19 fluid ounces of irrigation water. Thorough mixing of acid and water is essential prior to taking samples.

The amount of acid needed to lower a high pH to 6.5 can also be calculated by multiplying the milliequivalents (meq) of bicarbonate plus carbonate by one of the following factors:

- 7.0 for 75 percent phosphoric acid
- 3.2 for 65 percent sulfuric acid
- 10.5 for 67 percent nitric acid

If data for meq is not given but amounts of bicarbonates and/or carbonates are given as ppm of hardness, meq can be obtained by dividing the ppm by 50. If data is given as grains of hardness, the ppm is obtained by multiplying the grains of hardness by a factor of 17.

Low pH values of water need to be raised if used in low buffered media (sands, artificial soils) to avoid creating poor root environments. Ammonium fertilizers, especially ammonium sulfate, can cause serious reductions in pH in poorly buffered media. Modern hydroponic installations monitor the irrigation water. Low pH values can be raised by substituting calcium nitrate or potassium nitrate for some of the ammonium compounds. Some installations can introduce alkalis, such as ammonium hydroxide (aqua ammonia) or potassium hydroxide, to raise the pH. Ammonium hydroxide should not be used in closed areas where plants are growing because free ammonia in very low concentrations can be extremely toxic to plants.

Salts

The total dissolved salts in the water can be measured with conductivity tests that can easily be run by the grower. Conductivity is usually expressed as millimhos per centimeter (mmhos/cm) or the equivalent decisiemens per meter (dS/M).

Salinity may also be expressed as grains per gallon or tons of salt per acre foot of water. The grains per gallon can be converted to mmhos/cm by dividing grains by a factor of 37.45. Tons per acre foot can be converted to mmhos/cm by dividing tons by a factor of 0.87. also, tons of salt per acre foot of water is equal to ppm salts multiplied by 0.00136.

A conductivity of 1.0 mmhos/cm is about the level above which repeated applications of water for salt-sensitive crops can cause some reduction in yield. (An upper limit of about 0.7 mmhos/cm is more suitable for germination of many sensitive crops.) At a conductivity of 1.0, water will contain about 640 ppm of dissolved salts and will supply 1632 pounds of salts per acre foot of water. Amounts of applied water in acre feet for many crops are in the 0.5 to 1.0 range. Salts brought in by fertilizers, manures, and other additives, plus what is supplied by irrigation waters, may exceed the tolerances of some crops grown on many soils unless there is sufficient leaching of salts by rainfall or controlled irrigation. Salt damage is more severe in soils with low CEC, due to their lower moisture content.

The ideal conductivity range for salt-sensitive crops given above will still allow for the introduction of fertilizer. Slightly higher values than those given above can be used for salt-sensitive crops if fertilizers are not injected. Much higher values can be used for salt-tolerant crops, some of which can produce good crops with conductivity values in the 4-5 mmhos/cm range. Water with relatively high conductivity can be used for irrigation if there is sufficient rain between irrigations to prevent the accumulation of harmful amounts of salts (Table 8.5).

Coping with High Salts

Methods of coping with high salts in irrigation waters include: (1) preparing soil beds so that salts do not accumulate near plants, and (2) reducing the concentration of salt in the irrigation water. The concentration of salts can be reduced by either allowing for dilution by rainwater in ponds prior to water use and by the processes of ion exchange or reverse osmosis. Dilution by rainfall is

TABLE 8.5. Permissable number of irrigations between leaching rains using water containing varying amounts of salts and applied to crops with different salt tolerances

Irrigation water		No. of irrigations for crops having different salt tolerances		
Elec. cond.	Total salts	Good	Moderate	Poor
mmhos/cm	ppm			
1	640	—	15	7
2	1280	11	7	4
3	1920	7	5	2
4	2560	5	3	2
5	3200	4	2-3	1
6	3840	3	2	1
7	4480	2-3	1-2	0
8	5120	2	1	0

Source: Lunin, J., M. H. Gallatin, C. A. Bauer, and L. V. Wilcox. 1960. Brackish water for irrigation in humid regions. Bull. 213. USDA and Virginia Truck Exp't Sta. Agric. Inf.

possible only in certain regions. If used, evaporation must be restricted by the use of deep ponds and/or flotation of plastic beads or films. Removal of salts by either ion exchange or reverse osmosis is rather expensive and can only be considered for high-value crops. Even if used for such crops, water use may need to be restricted to (1) propagation of cuttings; (2) spraying of sensitive crops; and (3) the final rinsing of leaves after overhead irrigation with high-salt water.

Coping with Low Conductivity

Water with low conductivity can cause corrosion of irrigation equipment and have an adverse effect on water infiltration. The adverse effects, which increase in severity as the conductivity falls below 0.25 mmhos/cm, can be reduced by: (1) the addition of soluble fertilizers to the irrigation water; (2) passing the water over columns of gypsum stones; and (3) addition of fine particle gypsum (greater than 0.25mm) continuously during the irrigation period. The addition of fine gypsum is not practical for micro-irrigation systems. Conductivity is measured by conductivity meters as outlined in Chapter 3.

Sodium and Sodium Absorption Ratio (SAR)

The amount of sodium (Na) tolerated in irrigation waters is dependent on the sensitivity of plants to Na, and the degradation of soil by Na.

Many plants are sensitive to Na, while others respond favorably to Na application. Sensitive plants are listed in Table 8.4. Some plants, such as alfalfa, asparagus, barley, broccoli, brussels sprouts, carrot, cotton, flax, millet, salsify, tomato, and wheat, respond favorably to Na if K is low. Others, such as beet, celery, spinach, and turnip respond to Na even if K is found in higher concentrations.

Almond, apricot, avocado, citrus, and plum leaves rapidly absorb Na from leaf applications and are injured by very low concentrations in overhead irrigation water. Application of water that bypasses the leaves allows for safe use of concentrations five to ten times the level that would be harmful if applied directly to the leaves.

Application of water containing high levels of Na can cause physical problems for the soil, but the extent of problems is influenced by the type and amount of clay in the soil, and the amounts of Ca and Mg present in the water. The adverse effects are greater with increasing amounts of clay, particularly if the dominant clay mineral is montmorillonite rather than illite, kaolinite, or vermiculite.

The adverse effects of Na and their amelioration by Ca and Mg in irrigation water can be expressed by the use of the sodium absorption ratio (SAR), which is calculated from the following equation:

$$(3) \qquad SAR = \frac{Na}{\sqrt{\dfrac{Ca + Mg}{2}}}$$

where Na, Ca, and Mg are expressed as milliequivalents per liter (meq/1).

An SAR of less than 3.0 is considered as safe for most soils and crops. Problems increase as the SAR rises above 3.0, and become very serious at an SAR greater than 6.0. The amount of dissolved salts, however, has a bearing on the suitability of the different values, as depicted in Figure 8.1. Water falling into the four classes of Na hazard shown in this figure can be used under the conditions listed below.

1. Low Na hazard. Can be used on nearly all soils with little danger of building up excessive Na saturation, although the grower should watch for an accumulation of Na that can injure sensitive crops.
2. Medium Na hazard. Can be used for coarse-textured or organic soils with good permeability. May present some problems on fine-textured soils if leaching is limited and there is no gypsum in the soil.
3. High Na hazard. Can expect Na problems on most soils with continued irrigation, unless soils contain considerable gypsum and OM, and have good drainage.

4. Very high Na hazard. Usually unsatisfactory for irrigation except at low or medium conductivity, or if the soil contains large amounts of gypsum or the water is treated with gypsum.

Carbonates and Bicarbonates

Carbonates and bicarbonates increase water pH and with time the media pH as well. The change in media pH can be quite rapid in poorly buffered sands and artificial soils.

The presence of carbonates and bicarbonates in large quantities can cause immediate problems as several fertilizers are introduced, especially in small-orifice types of irrigation systems. Even without the addition of fertilizer, the presence of large quantities of carbonates and bicarbonates tend to increase clogging of irrigation lines.

The presence of large quantities of bicarbonates in water applied as overhead irrigation can also cause injury to plants. Most of the harm from bicarbonates is due to soil degradation over time, as Ca is depleted and replaced with Na.

Limiting Effects of High Concentrations

Fortunately, both carbonates and bicarbonates can be destroyed by the addition of acids to lower the pH as indicated earlier in this chapter. The loss of carbonates and bicarbonates as accelerated as the pH falls below 6.0, but corrosion of metal pipes and concrete ditches can be a problem at lower pH values.

The acid can be injected during the entire period of irrigation or limited to a short period toward the end of the irrigation cycle. Injection during the entire irrigation cycle eliminates all problems related to carbonates and bicarbonates. The late injection should be enough to prevent the plugging of lines and emitters, but extra gypsum will have to be added to the soil in time to prevent soil degradation caused by the untreated water.

The amount of acid needed to lower the pH of water is best determined by titration, but an estimate of the amounts of sulfuric

FIGURE 8.1. Classifying irrigation waters based on their Na and salinity contents

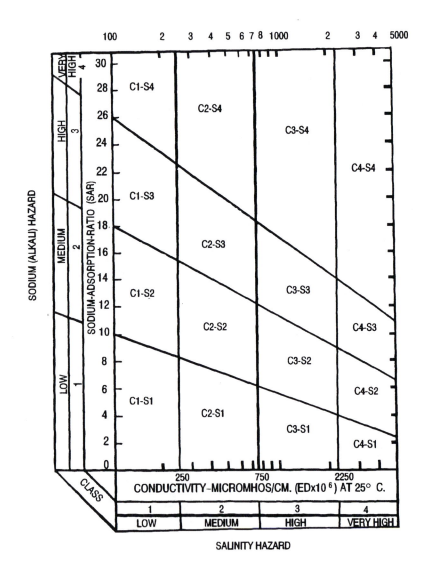

Source: Handbook 60, U.S. Salinity Laboratory.

acid needed to destroy 90 percent of the bicarbonates in an acre foot of water is given in Table 8.6.

Boron

The upper safe limit of boron (B) in irrigation waters is dependent on the tolerance of the crop to B, characteristics of the soil, and degree of leaching. The amount of B that can be toxic to plants is dependent on the amount of water applied, the type of soil, and the amount of leaching. Less B can be tolerated on the coarse textured soils low in OM, but B can easily be leached from these soils and is seldom a problem if there is enough leaching. The heavier soils, especially with high OM, tend to tie up B, allowing much greater amounts to be applied without causing injury. Such amounts can cause problems if soil pH is allowed to fall, making the B more available to plants.

Concentration Limits

Plants sensitive to B (see Table 8.4) should not receive water containing more than 0.5 ppm, but water with 2 ppm has been safely used for semitolerant plants, and as much as 4 ppm can be acceptable for highly tolerant plants. If water contains more than 0.1 ppm, B should not be used in fertilizers and foliar sprays used for sensitive crops.

TABLE 8.6. Quantities of sulfuric acid* required to neutralize 90% of bicarbonates in an acre foot of water

Bicarbonate	Acid required		
ppm	ppm	lb	gal
50	38	103	7
100	76	206	13
200	152	412	27
400	304	824	55

*95% sulfuric acid

Source: Stroehlein, J. L. and A. D. Halderman. 1975. Sulfuric acid for soil and water treatment. Arizona Agric-File Q-357.

Calcium

Water with very low or very high levels of calcium (Ca) can cause problems when applied as irrigation. Irrigating with waters containing less than 20 ppm of Ca tends to reduce the infiltration capacity of soils. Waters containing more than 100 ppm often tend to form precipitates, particularly as phosphates are injected into them.

Using Waters Low in Calcium

The harmful effects of low Ca waters are lessened by the introduction of fertilizers, particularly those containing calcium nitrate.

Using Waters with Elevated Calcium

Acidifying waters with nitric, hydrochloric, or citric acids reduces possible formation of precipitates. Injection of phosphates into waters containing Ca tends to produce precipitates unless the water is quite acidic.

A "jar test" to evaluate if precipitates are formed when phosphates are injected is always worthwhile. A 5-gallon container, glass cylinder, and sensitive balance are needed to perform the test. A pH meter or indicator papers will be helpful.

The test is run by adding ingredient(s) (injectable phosphates) in question to 5 gallons of the irrigation water in the same proportion that they will be injected into the lines. The water is stirred to dissolve the fertilizers and small quantities of fluid are transferred to glass cylinders or jars to observe whether precipitates are formed. A precipitate is indicated if the solution becomes cloudy or materials settle out within a 24-hour period.

Proportional amounts for the jar test can be calculated if one knows the quantity of water in which material(s) are to be injected. For example, if 50 pounds of a fertilizer is to be applied per acre by injection into an acre inch of water, the amount to dissolve in 5 gallons of water for testing is calculated from the proportion of fertilizer in 5 gallons to that to be dissolved in an acre inch of water (27,154 gals), or:

(4)

$$X:5::50:27,154,$$
$$\text{or } 27,154X = 250$$
$$X = 0.0092 \text{ pounds or } 4.18 \text{ grams } (0.0092 \times 454)$$

If precipitates are formed at normal water pH, the test must be repeated at lower pH values (5.5 and 4.5). Some chemicals that form precipitates in the pH range of 6.0 to 7.0 will be satisfactory if the water is acidified.

Chlorides

Injuries caused by excess chlorides in irrigation waters result from sensitivity to chlorides and damage from excess salts. Damage to many plants is much more severe if the water is applied by overhead irrigation.

Plants vary in their sensitivity to chlorides. Some sensitive plans are listed in Table 8.4. Alfalfa, cotton, and milo are moderately tolerant; sugar beets, barley, brussels sprouts, cabbage, cauliflower, corn, radish, spinach, and tomato are quite tolerant.

Leaf "burn" of many plants resulting from the overhead application of water high in chlorides results from the buildup of salts as the water evaporates. The presence of Na or other salts increases the amount of this type of damage.

Using Water with Higher Levels of Chloride

Growing plants tolerant to chlorides allows the use of chlorides to reduce damage from excess salts. Phytotoxicity caused by overhead irrigation is largely avoided by use of systems delivering the water directly to the soil. Even with overhead irrigation, phytotoxicity can be reduced somewhat if water with excess chloride is applied during periods of reduced evaporation (nighttime or the cool part of day and periods of little wind); by increasing speed of sprinkler heads to make at least 1 revolution per minute; and by positioning portable units away from the wind.

Fluorides

The upper safe limit of fluorides (F) in irrigation water is about 0.25 ppm for sensitive plants, but can be as much as 1 ppm for tolerant plants. A list of F-sensitive plants is provided in Table 8.4 (pp. 200-201).

Using Water with Elevated Fluoride

Slightly higher concentrations than the values given in Table 8.4 can be tolerated if the following conditions are met: (1) the soil pH is increased to 7.0; (2) transpiration is reduced by the use of windbreaks and partial shade; (3) overhead irrigation is avoided; and (4) the addition of F from other sources is minimized by using low-F sources of fertilizer and other soil amendments. Phosphates are the principal source of fertilizer F. Ordinary superphosphate and basic slag generally need to be eliminated if F in the irrigation water is a problem.

Fluorides can be removed from irrigation water by ion exchange or reverse osmosis processes or by treating the water with alumina or activated bauxite. Removal is an expensive process and probably is justified only for the production of some high-priced plants.

Iron

Problems related to excess iron (Fe) in irrigation waters arise because of two different effects. First, iron present in water applied by overhead irrigation tends to precipitate on leaves, causing unsightly stains. Second, relatively small amounts of iron can cause serious blockage of small orifice emitters as bacteria utilize the iron to form sludges.

The economic damage from deposits on leaves is of minor consequence, except for foliate or flowers where the deposits detract from the plant's visual appeal. As little as 0.1 ppm of Fe can be troublesome if the water contains considerable bicarbonates or carbonates (pH greater than 7.0).

Using Water with Elevated Iron

The problem of staining is mitigated by acidifying the water to a pH of 6.0 or less, or by adding chelating agents, such as hexametaphosphates (Quadrophos or Calgon) to keep the iron in solution.

Unsightly Fe deposits on ornamental plants can be removed by spraying leaves with a 5 percent oxalic acid solution (6.7 ounces or 16 tablespoons in a gallon of water) and then rinsing before the spray dries. The spray can be phytotoxic and should be tried on a few plants before using it on a large scale.

The method of controlling clogging problems caused by excess Fe will depend on the amount of Fe present and whether large amounts of manganese (Mn) are also a problem. Small amounts of Fe can be handled by the addition of chlorine into the water before the filters, but larger amounts (greater than 0.4 ppm Fe) will require aeration and settling ponds in addition to chlorination. Aerating the water will help oxidize much of the Fe, which is relatively insoluble and settles out in the reservoir. Aeration can be increased by allowing the water to tumble over rocks as it enters the reservoir, or by pumping the water into the air and allowing it to fall back into the reservoir.

The use of chlorine to precipitate Fe is not a good option if Mn concentrations are greater than about 0.2 ppm. Unlike Fe, which tends to precipitate quickly as chlorine is added, Mn tends to precipitate slowly, allowing it to bypass the filter and settle out in the lines. If both Mn and Fe are high, chlorine additions have to be postponed until the end of the irrigation, at which time sufficient acid is also added to keep the pH at about 5.5. Avoid the addition of acids with hypochlorites.

Magnesium

The higher levels of magnesium (Mg) in the tolerable range are satisfactory provided salts are not excessive and materials such as phosphates can be injected without causing precipitates. The latter can be evaluated by jar tests. If precipitates are formed, satisfactory use is still probable by injecting acids to lower pH.

Nitrogen

The natural occurrence of nitrogen (N) in irrigation water would appear to be a "freebie," which it often is, but its presence can cause problems at times. More than about 5 ppm can stimulate the growth of algae in ponds, increasing the problem of orifice plugging. Amounts greater than 30 ppm can delay the maturity of some crops and decrease the sugar content of fruits. The presence of large quantities of N probably is beneficial for most crops during the early stages of growth, but usually N must be reduced as the crop approaches maturity.

Using Water with Elevated Nitrogen

In most situations, the presence of N will be beneficial if the amount provided by irrigation is considered and deducted from the fertilizer recommendation. For example, a concentration of 25 ppm will supply about 68 pounds of N in an acre foot of water. Such amounts supply enough N to grow some crops on certain soils.

Sulfides

The presence of sulfides (S) in irrigation water in amounts greater than 0.1 ppm can lead to the production of slimes that clog irrigation systems. Insoluble iron sulfides that clog systems can be formed by the interaction of Fe with sulfides naturally present in the water, or by the introduction of Fe in added fertilizers.

Using Water with Elevated Sulfides

An excess of sulfides can be handled by the use of chlorine, using much the same procedure described for dealing with Fe.

Chapter 9

Measuring Flow Rates and Volume

Flow rates and volume measurements enable the grower to determine whether plants are receiving scheduled volumes. It is important for the grower to know how much water is being applied in a given time period in order to plan the application of fertilizer or other chemicals through the system and also to effectively recharge the soil.

Flow decline as indicated by repeated readings of flow meters can indicate problems with pumping stations or clogging. Sudden increases in flow rate could be an indicator of system leakage or pipeline breakage.

METHODS

Meters

Flow meters can be installed in pipelines to measure the volume of water applied. They can be placed at the head of the farm to show total volume consumed or at the head of individual fields. The latter is preferred, but all types of installation need to provide data as to instantaneous and cumulative flow. Metering devices which shut off the flow of water after a predetermined amount is delivered are an effective means of measuring the volume of water used.

Measuring Sprinkler Delivery Rate

The delivery rate of nozzles, as noted by the manufacturer or determined by measurement, can be used to estimate delivered volumes, but discharge rates in the field are a more accurate method of determining the volume delivered. The average discharge rate of several sprinklers is determined by directing each nozzle flow into a 10-50 gallon tank for a period of ten minutes and then measuring the volume of water collected. The rate (R) of water applied can be calculated from the equation:

(1)
$$R = \frac{SR \times 96.3}{DS \times DL}$$

where R = inches per hour (in/hr)
    SR = average sprinkler rate in gallons per minute (gpm)
    DS = distance between sprinklers in feet
    DL = distance between lines in feet

Measuring Drip Irrigation Delivery Rate

The delivery rate of drip systems can be estimated in a similar manner, except that it is desirable to make the collections in small containers for point emitters, or in troughs for systems with porous tubing. The contents are transferred to 100 milliliter graduated cylinders to measure the volume in milliliters (ml). It may be necessary to remove some soil under each emitter or a length of porous tubing equal to the trough length so that the collection containers can be placed underneath without lifting the pipe. The milliliters per minute (ml/min) can be readily converted to gallons per minute (gpm) from the data in Table 9.1. Equation #1 above can be used for estimating inches per hour delivered by point emitters, with DS being the distance between point emitters or the length of the trough used for collection from porous tubing.

Measuring Open Channel Delivery Rate

The volume of water passing through open channels can be calculated from the equation:

(2)
$$V = A \times MV$$

Where V = volume in cubic meters (cubic meter = 264.17 gallons)
    A = cross section area of flow in square feet
    MV = mean velocity of flow through the cross-section

The cross-section area (A) of smooth-sided channels can be calculated, but several points along channels with nonuniform sides must be averaged to obtain their area. The mean velocity (MV) can be taken from a meter measuring the current, but this may not be available for most growers. A simple (but only approximate) method of estimating MV is to measure the movement of a float over a standard straight length of the channel for 150 feet. The MV in feet per second for smooth-lined ditches will be 0.8 the surface velocity measured by the float; the MV of rough ditches is 0.6 the surface velocity.*

Measuring Flow from Horizontal Pipes (Trajectory Method)

Flow rate from full pipes can be measured by determining the horizontal distance X (in inches) parallel to the centerline of the pipe needed for the trajectory to fall a vertical distance Y of either 6 inches or 12 inches. If the pipe is full, the measurements are made on the upper surface of the jet; if partially full, the measurements are made at the center of the jet. Well yield is equal to the horizontal distance of the trajectory (X) multiplied by a factor from Table 9.2. For example, if the length of the trajectory is 19 inches when the height of the trajectory is 12 inches for an 8-inch pipe, the factor is 52.9. This factor multiplied by 19, the trajectory distance, gives a flow rate of 1005 gpm.

TABLE 9.1. Conversion of milliliters per minute to gallons per minute

ml/min	gpm	ml/min	gpm
45	0.0118	75	0.0198
50	0.0132	80	0.0212
55	0.0145	85	0.0225
60	0.0158	90	0.0238
65	0.0172	95	0.0250
70	0.0185	100	0.0263

*More complete methods of measuring MV as well as measuring water flow over several different types of weirs and through flumes are covered in some detail in Hillel (1987) and Scott and Houston (1981). See Additional Reading at the end of this chapter.

Measuring Flow Through Siphons

The amount of water flowing through siphons can be determined from Figure 9.1. The head is the vertical distance from the surface of the water in the inflow ditch to the surface of the water in the outflow ditch. The flow rate is obtained by finding the head distance in Figure 9.1 and drawing a line horizontally until it reaches the appropriate line for the internal diameter (inches) of the siphon. A perpendicular line drawn from this intersection point to the x-axis gives the flow rate in gpm.

UNIFORMITY OF WATER APPLICATION

Lack of uniform water application can lessen the efficiency of applied water. Solutions to this problem are complicated as fertiliz-

TABLE 9.2. Factors for converting trajectory height of water flowing from pipes of several dimensions to gallons delivered per minute

Pipe diameter in.	Factors for height trajectories (Y) of	
	6 in.	12 in.
2	5.02	3.52
3	11.13	7.77
4	17.18	13.4
6	43.7	30.6
8	76.0	52.9
10	120.0	83.5
12	173.0	120.0

Note: Directions for measuring "Y" and other methods for calculating water flow can be obtained from Scott, V. H., and C. E. Houston. 1981. *Measuring Irrigation Water.* Leaflet 2956. Division of Agricultural Science, University of California, Berkeley, CA 94720.

Source: Schwab, D. *Irrigation Water Management.* OSU Extension Facts #1502. Cooperative Extension Service, Division of Agriculture, Oklahoma State University, Stillwater, OK.

FIGURE 9.1. Calculating the rate of flow through small (A) and large (B) siphons

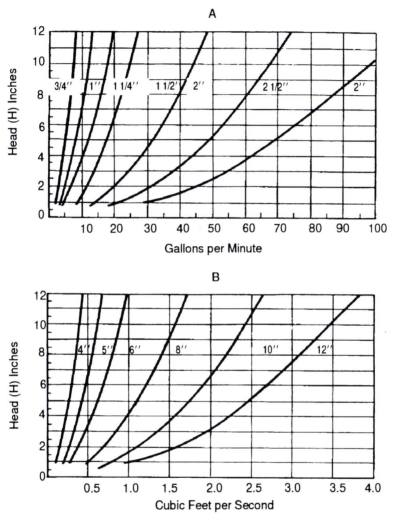

Note: To measure rate of flow start at the left, moving horizontally from the appropriate "head in inches" until reaching the size of pipe or siphon. Move downward to find the rate of flow in gallons per minute. The head is calculated as difference in elevation between water in ditch and water in the field.

Source: Scott, V. H. and C. E. Houston. 1981. *Measuring Irrigation Water.* Leaflet 2956. Division of Agricultural Science, University of California, Berkeley, CA 94720.

ers are injected into the water, as plant growth will then not only reflect variation in available water, but in nutrients as well.

Measuring Uniformity

Uniformity is measured by collecting water as indicated in the sections on Sprinkler Delivery Rate and Drip Irrigation Delivery Rate from a number of sites in a given field. But rather than measuring the amount delivered, measure the time (using a stopwatch with a second hand) that it takes to fill a standard container. The container can be as large as 5 gallons for sprinkler heads or as small as a soda bottle for drip emitters.

The uniformity of water delivery can be determined by timing the collection in a large number of places. It has been estimated that about 18 locations scattered uniformly over a field are necessary to test the uniformity of drip irrigation. The three highest values are added together and this sum is divided into the sum of the three lowest values and expressed as a percentage. The acceptability of a given level of uniformity is indicated in the following table.

Degree of uniformity	Relationship of low to high values Top three values ÷ Lowest three values as a percentage
Excellent	90-100
Very good	80-90
Fair	70-80
Poor	60-70
Unacceptable	<60

TIME REQUIRED FOR INJECTING CHEMICALS

Timing the injection of fertilizer or other chemicals into the irrigation system enables the grower to fully apply materials without unduly diluting their concentration. Proper timing of injected materials into drip irrigation systems also allows for proper placement since the amount of water following injection will affect the movement of materials away from emitters.

Procedure

Before attempting to determine the time necessary to inject a given material, the system must be applying water uniformly as outlined in the sections on Sprinkler Delivery Rate and Drip Irrigation Delivery Rate. Ideally, uniformity ought to be on the excellent range or at bare minimum in the very good range.

The timing is determined from the interval needed for an indicator solution consisting of a dye or detergent to move from the injector to the last emitter. The following steps outline the procedure:

1. Operate the system until all emitters are working and balanced at full pressure.
2. Make up a solution of dye or detergent in a known volume (full tank).
3. Inject the solution and start timing.
4. Monitor the appearance of the solution at the closest and farthest points from the source of water.
5. Calculate the volume of solution injected by measuring the amount of solution remaining.
6. Monitor the time of final appearance of the solution at the farthest emitter.
7. The time interval from the start of injection until the first appearance at the last emitter is the minimum time required for injection. The minimum injection rate can be determined by dividing the total gallons in the tank by the minimum time of injection. The time interval from the start of injection until the appearance of the final solution at the farthest emitter can be used as a check against these calculations. If partial volume cannot be estimated (step 5 above), the time interval between the injection of the solution and its final appearance at the last emitter serves as the amount of time needed to inject a given amount of diluted solution.

Additional Reading

Ayers, R. S. and D. W. Westcott. 1976. *Water Quality for Agriculture*. Irrigation and Drainage Paper #29, FAO, 00100, Rome, Italy.

Boswell, M. J. 1990. *Micro-irrigation Design Manual*. James Hardie Irrigation, Inc. El Cajon, CA 92020.

Bucks, D. A., F. S. Nakayama, and R. G. Gilbert. 1979. Trickle Irrigation Water Quality and Preventative Maintenance. *Agricultural Water Management* 2: 149-162.

Goldhamer, D. A. and R. L. Snyder (ed.). 1989. *Irrigation Scheduling*. Division of Agriculture and Natural Resources Publication 21454. University of California, Oakland, CA 94608-1230.

Hillel, D. 1987. *The Efficient Use of Water in Irrigation*. World Bank Technical Paper 64. World Bank, Washington, DC 20433.

Kano, Y. 1990. Measurement of soil moisture potential. In *Measurement Techniques in Plant Science*. Y. Hashimoto, P. J. Kramer, H. Nonami, and B. R. Strain (eds.). Academic Press, Inc., San Diego, CA 92101.

Proceedings of the Third Annual Micro Irrigation Workshop and Mini-Trade Show. 1990. Donald Pitts, Coordinator. Southwest Florida Research and Education Center, Immokalee, FL.

Sawyer, G. C. and T. W. Oswalt. 1984. Injection of agricultural chemicals into micro-sprinkler systems. *The Citrus Industry*, (October) 65(10): 32-37.

Scott, V. H. and C. E. Houston. 1981. *Measuring Irrigation Water*. Leaflet 2956. Division of Agricultural Science, University of California, Davis, CA.

SECTION IV:
DIAGNOSTIC PROCEDURES
FOR MAXIMIZING PEST CONTROL

It has been estimated that about one-fourth of all crop production worldwide is lost annually to the following groups of pests: (1) insects and mites; (2) bacteria and fungi; (3) viruses, viroids, and mycoplasmas; (4) nematodes; and (5) weeds. The losses could be reduced by the more efficient application of control methods if diagnostic procedures were used more frequently to: (1) identify the pests; (2) count their numbers; (3) apply effective materials when threshold numbers are present or conditions are favorable for a pest outbreak; and (4) apply the corrective treatment at the right dosage.

Chapter 10

Identifying and Counting Pests

Identification of the pest is the first line of defense, as it defines the method of control. In many cases, identification is a very simple procedure, as both the pest and its damage are visible. If the pest is not immediately recognizable, it can be compared with illustrations or photographs. There are a number of books (some of which are listed in Additional Reading at the end of this section) that contain photos or illustrations of various pests. Additional lists are available through state Extension Services or from various pesticide manufacturers.

Often, the particular pest causing the problem is not readily apparent because it is too small to be seen or it may move rapidly from areas of damage. Pests that cannot be seen readily are bacteria, fungi, nematodes, viruses, viroids, and mycoplasmas, all of which are either microscopic or submicroscopic. Most mites, many of which are not easily visible to the naked eye, can be seen through a hand lens of 10–20 power. A number of sucking insects, such as aphids, thrips, treehoppers, and whiteflies, move about rapidly and often are not identified as the source of plant damage until their numbers are relatively large. In addition to sucking plant juices, these pests cause damage by transmitting viruses, viroids, and mycoplasmas.

Counting the numbers provides information as to the need for treatment and when treatment should begin. The manner in which the pest is identified and counted depends largely on the type of pest. The procedures are briefly described under the separate categories below.

INSECTS AND MITES

Most chewing insects can easily be identified because they are readily associated with plant damage. If identity is not immediately established, the insect can be compared with illustrations or photos, or collected and sent to Extension Services for identification.

Some knowledge of the pest, its feeding habits, and its stages of growth aids in collection and identification.

Insects

Insects go through metamorphosis or changes as they develop from egg to adult. The changes can be rather limited, with the young appearing much as an adult except for size, absence of wings, and external reproductive organs, as is the case with grasshoppers, plant bugs, and aphids. Or the metamorphosis can be complete, wherein the egg hatches into a larva or wormlike creature. The eggs of butterflies and moths become caterpillars; those of beetles and wasps hatch into grubs; and those of flies into maggots. The larval stage is followed by a resting or pupal stage, with a number of pupae enclosed in a protected cocoon of silk, earth, or other covering. The insect may overwinter in the larval stage, but will finally emerge as a winged adult. This type of complete metamorphosis is typical of butterflies, moths, beetles, ants, bees, wasps, and flies.

Collection and Counting

The process of collection and counting is most effectively accomplished by a combination of scouting and trapping.

Scouting. Scouting involves a careful examination of plants on a regular basis. Plants may be examined randomly if there are no indications of insect damage. Often, the presence of insect damage may be indicated by abnormal color, size, vigor, or leaf shape, and plants exhibiting such problems ought to be examined first. In the absence of abnormal plant characteristics, a more random examination is often satisfactory, although the ends of benches or rows should be examined first.

Plants need to be examined carefully. Starting at the top, examine both upper and lower surfaces of leaves, stems, buds, blossoms, and fruits, always looking for the presence of eggs, nymphs, larvae, and adults. Detection of eggs and nymphs is greatly aided by the use of a hand lens. The numbers of insects are counted and the affected plants are marked for future examinations.

In some cases, roots also may need to be examined. Most potted plants can be removed from the pot and examined for the presence of root aphids and grubs. Square foot samples of sod and soil can be removed and examined for insects, many of which live at the soil-thatch interface.

Scouting for insects can be aided by the use of sweeps and ground cloths. Sweeps are especially helpful in collecting and counting insects that move readily from place to place. Sweep nets can be purchased or made up by attaching a 12- or 15-inch diameter cone of muslin to a sturdy handle that is 3-4 feet long. A ground cloth can consist of white or tan canvas about 30 inches square that is stapled or sewn on two pieces of lightweight wood (dowel rods or broom handles) about 5 inches longer than the canvas.

Insects are caught up in a sweep net as it is moved quickly near the plants in a sweeping motion. After a set of sweeps, the net is swung quickly back and forth to force insects to the bottom of the sweep net. The net is grabbed and closed 6-8 inches from the bottom to prevent escape before the insects are counted. With large numbers of insects, some can be allowed to escape after they are counted in order to facilitate counting the remainder. The total number of insects, as well as the number of sweeps, should be recorded.

Insects can be collected in a drop cloth by shaking or striking a branch with leaves while the ground cloth is held underneath it. The cloth is kept rolled up while approaching a sampling site being careful not to disturb plants. After the cloth is placed unrolled on the ground (between rows if row-cropped), the cloth is unwrapped quickly, held in place with the knees, and the plants are shaken and the insects counted quickly. In counting, care must be taken not to miss caterpillars or beetles still clinging to fallen leaves. The number and kind of insects, as well as the number of plants shaken, should be recorded.

Collection of insects by sweep net or ground cloth is best done on sunny, low humidity days. Sampling sites, about a foot ahead of the person doing the sample, need to be approached carefully so as not to disturb the insects before shaking the plant. Collection and counting must be repeated in at least several places in a field in order to minimize variability.

Trapping. Insects may also be collected by traps. Traps suitable for such purposes are available in three different types. The visual type reflects certain light wavelengths in a manner that resembles a leaf or fruit. Bait traps contain certain essences resembling the odors of plants or ripe fruit consumed by the insect. Pheromone traps contain artificial chemicals which mimic the odors used by female insects to attract mates. Traps contain sticky substances that hold insects as they alight or are built so that it is difficult for the insect to escape once it has crawled in.

There are commercial traps available for collecting apple maggots, Japanese beetles, peach borers, codling moths, and cherry fruit flies. Yellow traps lure aphids, gnats, black flies, leafhoppers, moths, and whiteflies away from plants and capture them as they land. Red sphere traps, baited with apple scents, are effective in collecting apple maggot flies. There are a number of pheromones with which traps can be baited and are capable of attracting more than 50 different types of insects. In addition to these traps, yellow sticky cards or traps can catch aphids, black flies, whiteflies, leafhoppers, gnats, and moths. White sticky cards are better suited to trapping flea beetles and tarnished plant bugs. Blue sticky traps are primarily used to collect thrips. If only one color is used, yellow is preferred because of its attractiveness to the widest variety of insects.

A simple visual trap can be made from small rectangular boards, poster boards, plastic plates, or from 3-pound coffee can lids. The materials are painted with Rust-Oleum 659 yellow and coated with SAE-90 oil, cooking oil, petroleum jelly, or a commercial preparation, such as Pestick.

Prepared traps, some with lures and pheromones, can be purchased from a number of companies, several of which are listed in Table 10.1.

The traps are usually placed at about the height of the canopy. In greenhouses, where they are most effective, they can be hung by wooden clothespins glued back-to-back and clipped on to a bamboo stake poked into a pot, or they can be attached to mist lines, tent support wires, or cord suspended from the greenhouse. In the field, the traps are usually placed at canopy height on stakes placed in the row. On golf courses and lawns, a pitfall trap consisting of a plastic cup (16-20 ounces) with holes punched in the bottom is placed with

TABLE 10.1. Companies supplying insect traps

---

1. Agri-Sense, 4230 W. Swift Ave., Fresno, CA 93722
2. Arbico, PO Box 427 CRB, Tucson, AZ 85738-1247
3. Consep Membranes, Inc., PO Box 6059, Bend, OR 97708
4. Gardens Alive, PO Box 149, Sunman, IN 47041
5. Great Lakes IPM, 10220 Church Rd., NE, Vestaburg, MI 48891
6. Ladd Research Industries Inc., PO Box 1005, Burlington, VT 05402
7. Necessary Trading Co., New Castle, VA 24127
8. Olson Products, Inc., PO Box 1043, Medina, OH 44253
9. Pest Management Supply Inc., PO Box 938, Amherst, MA 01004
10. Seabright, Inc., 4026 Harlan St., Emoryville, CA 94068
11. Scentry, Inc., PO Box 426, Buckeye, AZ 85326
12. State College Labs., 500 Spring Ridge Dr., Reading, PA 19612
13. The Tanglefoot Co., 314 Straight Ave., SW, Grand Rapids, MI 49504
14. Trece, Inc., PO Box 6278, Salinas, CA 93912

---

its top at ground level in a hole made by a standard 4.5-inch golf course cup cutter.

The number of traps needed varies with plants, pest, and location. In the greenhouse, one to four traps are placed per 100 square feet, but in the field the numbers per 100 square feet are greatly reduced.

The traps must be checked at each scouting interval, or at least once a week, at which time insects are counted and identified. Examination and counting are aided by using a hand lens to enhance the insects' characteristics and a small forceps to handle the entrapped insect. Traps need to be changed at least once a week or when they become covered enough to make counting difficult. Hanging traps for a few hours and making a count can help determine the effectiveness of recent insecticide applications.

Automating Bug Counting

Remote counting of certain insects is possible and can be helpful in evaluating insect problems in remote fields and/or for several greenhouses. Remote counting relies on the use of an effective pheromone or bait. The insect, attracted by the bait or pheromone, falls into a chamber where it is killed by an insecticide. The insect falls through a funnel and drops on a sensor plate, which triggers the count. Numbers for each date can be transferred by telemetry to office computers where they can be displayed and stored for future reference. Software is available that can plot insect counts over time and sound a warning whenever counts exceed threshold limits. Such data can greatly aid in evaluating the need for releasing predators or starting pesticide applications. A suitable system of counting and telemetry is available from Automata, Inc., Grass Valley, CA 95945-8816.

Identification

If identification of the collected insect is not possible, unidentified insects can often be sent to the State Extension Services for identification. Integrated Pest Management (IPM) Services and some commercial laboratories can also help make the identification. Directions for shipping specimens can be obtained by contacting these services. Local county agents can provide directions for shipping to Extension Services. If directions are not available, satisfactory specimens will usually reach their destination if (1) soft-bodied insects (ants, aphids, grubs, maggots, and thrips) are placed in 70-percent alcohol or 100-proof liquor, and (2) hard-bodied insects (beetles, flies, moths, and wasps) are first killed by rubbing alcohol. The soft-bodied insects should be submitted in a plastic vial with the lid tightly secured; the hard-bodied insects are shipped in dry crush-proof containers. No live insects can be submitted, nor is it practical to place loose specimens in an envelope. The properly prepared specimens need to be carefully packed in a tight container. Before closing, include an information sheet that provides the grower's name, address, telephone number, and a description of where specimens were found, the numbers present, and the extent of damage.

Mites

Although usually considered with insects, mites belong to the class Arachnida. Development, although it is different for different species, follows a typical life cycle. The adult female lays several hundred eggs in her lifetime. These are attached to silk webbing and tend to hatch in a few days, going through three molt stages before reaching the adult stage. There are several kinds of mites that damage crops.

The *spider mite* is about 1/60th of an inch in length and may be white, green, yellow, purple, black, or red. The harm done by spider mites is due to the sucking of juices primarily from the underside of the leaf. In hot dry weather, the damage will turn the leaves blotchy with pale-yellow, reddish brown spots. Leaves become sickly, die and fall prematurely. The *two-spotted* mite is characterized by two spots on its back. Small webs that help protect it may be present. The webs cause no harm to plants, but can be unsightly on flowers and ornamentals.

The *cyclamen mite*, which is a serious pest in greenhouses, is microscopic in size and translucent. Its presence is favored by temperatures lower than 60°F and a relative humidity greater than 80 percent. It has an average life span of about two weeks, although females live on for three to four weeks. They infest new leaf and blossom buds, curling the leaves and distorting blossoms, which may prevent flower production or cause premature abscission.

The *citrus rust mite*, about 0.005 inches long, feeds on both sides of citrus leaves and green fruit, leaving rusty, rough patches on the fruit. Mite feeding tends to have an adverse effect on tree vigor, often leading to the decline of the tree.

Collection and Identification

Collection of spider mites for identification is simplified by shaking leaves allowing debris to fall on white paper. The moving mites can be readily examined by hand lens. Because of its extremely small size, the presence of cyclamen mites must be confirmed by examining tissue under a microscope. Rust mites can be readily identified by examining affected leaves or russeted fruit with a hand lens.

Identification is made by comparing collected specimens with photos and illustrations available from many state extension services and insecticide manufacturers. Scouting personnel providing IPM services are usually well qualified to make the identification. If identification is not possible by these means, specimens can be sent to certain state extension or private laboratories. Sending portions of plants (russeted fruits or leaves) aids in detecting rust mites. If possible, entire plants should be submitted for identification of cyclamen mites. Follow general directions for preparing and shipping samples as given in steps 1, 3, 4, 7, and 8 in the next section, except that special care must be taken in the handling and packing of plant portions to avoid dislodging mites. Placing leaves or fruit in tight plastic bags before packing aids in their retention.

BACTERIA AND FUNGI

The collection and identification of bacteria and fungi that cause diseases are much more difficult than is the case with insects because the former are microscopic in size.

Bacteria

Identifying Bacteria

Tentative identification of bacterial diseases can be made based on the type of damage incurred: (1) *wilts* or *blights* caused by the plugging of conducting cells; (2) *spots* formed by the destruction of leaf or fruit cells; (3) *rots* due to the breakdown of fruits, rhizomes, storage roots, tubers, and harvested produce; (4) *galls* produced by the stimulation of tissue growth; and (5) *stunts* due to the suppression of cell growth. Frequently, symptoms exhibited can be matched with illustrations, photographs, and/or descriptions of disease to allow identification.

Positive identification is generally accomplished by microscopic examination. At times, a more positive identification is needed. This can be accomplished by means of reacting with dyes, enzymatic action, pathogenicity studies and noting their susceptibility to certain viruses or bacteriophages.

Such specialized identification is usually beyond the capabilities of most growers, but is often available through state extension services or private laboratories. The state extension services can be contacted through county agents. Directions for preparing the samples and their shipment so that samples arrive in satisfactory condition for evaluation can often be obtained from the laboratory making the tests. If not available, generalized directions given below probably will be satisfactory.

Submitting Samples for Identification

1. Provide an information sheet which supplies the name and variety of plant affected, prevalence and severity of the problem, location, fertilization and spray program, and cropping history, along with name, address, and telephone number of the grower. The information sheet should be placed in an envelope in a plastic bag inside the container.
2. Supply adequate samples that have not deteriorated. Partially diseased samples that reflect several stages of the disease are preferred.
3. Submit entire small plants with as much of the root system included as possible. Wash soil from the roots and wrap the entire root system in wet newspapers or toweling. Place the wrapped root system in a plastic bag, and secure the bag above the roots with rubber bands, ties, or cord. Place the entire plant in a bag, using a paper bag if any soft rotted tissue is present.
4. Submit parts of large plants. Again, samples showing various stages of disease are preferred. For leaf spot problems, submit a portion of cut stem with the leaves attached, wrap the cut part of the stem in wet wrap as above, and place the entire sample in a plastic bag.
5. Submit several branch tips (0.25-0.50 inches in diameter and 6-8 inches long) with leaves in a bag for evaluation of cankers from trees or shrubs.
6. Submit large leaves with spots, and place them in plastic bags before shipping.
7. Pack in a sturdy container and seal tightly.
8. Rush samples to a laboratory or clinic. If samples cannot be sent immediately, refrigerate samples until they can be sent.

Predicting Bacterial Disease Outbreaks

Unfortunately, if treatment is delayed until disease symptoms are sufficiently advanced for identification, much crop damage can occur and it is often difficult to control the disease once it has started. Starting a preventative program before symptoms appear can greatly reduce losses from some diseases.

A number of bacterial diseases in the field are favored by warm, moist conditions and high relative humidity. The bacteria are readily moved by splashing water. Entry occurs through natural plant openings such as stomata, or through breaks and cuts caused by pruning, rough handling, and wind damage. Since the organism is motile, free water is essential for entry to occur. Weather conditions that favor rapid disease development form the basis of computer programs capable of predicting serious outbreaks of disease. Some of the programs that use weather conditions to predict the onset of bacterial inflections are listed in the section on fungi that follows.

Fungi

A description of fungi and of fungal development can aid in their identification. A fungus starts from the germination of *spores*, which have a function similar to seeds in higher plants. They are usually produced on aerial structures that are well differentiated from the mycelium in the substratum. Most fungi produce nonmotile spores in tremendous numbers. They grow or germinate by pushing out a filament or *hypha* from a thin part of the wall. The hypha tube continues to grow and becomes the *mycelium*. The hyphae and mycelia are the vegetative stages of the fungus and grow in a medium by branching. The medium can be either decaying or live tissue. Infection of live tissue may occur by means of direct penetration of undamaged tissue, through natural openings (stomata), or through wounds.

Ideal germination conditions for most spores are temperatures from about 60-70°F and the presence of free water. A few powdery mildews can germinate under conditions of high relative humidity (97-100 percent). Under ideal conditions of moisture and temperature, fungal diseases can spread very rapidly. A few days of warm, sultry weather can transform a minor infection into one that leads to almost complete devastation, which underscores the necessity of taking protective measures before such conditions arise.

Some fungi enter a resting or dormant stage and form *sclerotia*, closely compacted mycelia that vary in size from a pinhead to several inches in diameter. As sclerotia, they are resistant to desication and other adverse conditions. When conditions are favorable, sclerotia germinate to continue the vegetative state or produce spores.

Identification of Fungi

As with bacteria, tentative identification can be made on the basis of symptoms by comparison with photos and illustrations. A number of pictorial guides are available for identifying plant diseases. Some of the best material for identifying diseases caused by bacteria as well as fungi is contained in a series of monographs dedicated to diseases of individual crops published by the American Phytopathological Society, St. Paul, MN 55121-2097.

Submitting Samples for Identification

Positive identification will at times require examination with a microscope or more-involved testing. Unless one is trained and facilities are available for such tests, samples will have to be forwarded to qualified laboratories capable of making a positive identification. Some commercial and state extension laboratories employ pathologists who can make these identifications. As was suggested in the earlier discussion about submitting samples for bacterial identification, following instructions as to sampling, packaging, and shipment will help to provide more meaningful results.

Test Kits

Kits that allow growers to make on-farm identification of several fungal diseases have recently been made available. Evaluation of the disease proceeds with the preparation of an extract by grinding roots, stems, or leaf samples and suspending the extract in a special extraction solution. The extract is applied by an extraction solution bottle that filters out particulate matter leaving a clear liquid that is added to three wells. Several test solutions are added to the wells which produce colors that form the basis for disease identification. The liquid in the "positive control" well develops a blue color if the test was performed properly; the liquid in the "pathogen indicator" will develop a blue color if the sample contains a pathogen; the "negative control" should remain colorless. A darker blue color in the pathogen indicator than in the negative control indicates a positive test. Some of the diseases of several crops detected by these kits are listed in Table 10.2.

Predicting the Onset of Disease

Integrated Pest Management (IPM), with its routine scouting services, can be effective in determining the first appearance of disease. If controls have not been initiated by that time, effective spraying with appropriate materials needs to be started at once. But starting control programs before symptoms appear can greatly reduce losses from a number of fungal and bacterial diseases. Savings in such spray programs can be effected if treatments are delayed until conditions of temperature, moisture, and the presence of the disease in other areas warrant their start. Some extension services maintain a monitoring service that will alert the grower to the need to start a particular program.

Forecasting Service. So closely is the development of certain pests related to environmental conditions, that fairly accurate forecasts for starting control measures can be made from certain atmospheric data.

TABLE 10.2. Crop diseases detected by Alert* detection kits

Phytophthora	Pythium	Rhizoctonia	Sclerotinia
alfalfa	asparagus	chrysanthemum	dry bean
chrysanthemum	chrysanthemum	hibiscus	snap bean
hibiscus	cotton	Ilex	celery
Ilex	hibiscus	juniper	peanut
juniper	Ilex	potato	
pepper	juniper	rice	
potato	peanut	schefflera	
raspberry	schefflera	soybean	
schefflera	soybean		
soybean			

*Alert On-Site Disease Detection Kits, Agri-Diagnostics Association, 2611 Branch Pike, Cinnaminson, NJ 08077.

The ability to predict the onset of disease based on weather data is being put to practical use. In the U.K., the application of fungicides to control late blight on potatoes is recommended whenever weather conditions favoring sporulation and rapid spread of the pathogen are prevalent for more than two days. Conditions favoring the disease are night temperatures above 50°F and relative humidity greater than 90 percent. Tom-Cast, a weather-based prediction system for tomato diseases in Ohio, uses four locations and grower-supplied weather data (temperature and leaf wetness values) for recommending fungicidal applications.

The presence of high-risk conditions for some cereal diseases have also been used in the U.K. as a basis for recommending fungicide applications. The high-risk period starts with a high-risk day but ends either when there is not more than one factor favoring high-risk days, or on the third consecutive day that only two factors favoring high-risk days are present. High-risk days for powdery mildew of barley are initiated whenever the following occur simultaneously:

1. Maximum temperature greater than 59°F.
2. More than five total sunshine hours.
3. Rainfall less than 0.04 inches.
4. A total daily run of wind greater than 153 miles.

Several turf diseases can be predicted from atmospheric data. Damage from Pythium blight (*Pythium spp.*) can be expected if in a 24-hour period: (1) the maximum air temperature is greater than 82°F; or (2) the minimum air temperature is greater than 68°F and there are three hours of relative humidity greater than 90 percent; or leaf wetness greater than 9. Brown patch (*Rhizoctonia solani*) can become a problem when: (1) the air temperature exceeds 61°F and the average temperature is greater than 70°F; (2) the relative humidity remains higher than 95 percent for at least 12 hours; (3) soil temperatures are greater than 68°F and average greater than 70°F; and (4) the rainfall is greater than 0.22 inches. Dollar spot (*Sclerotinia homoeocarpa*) outbreaks can occur whenever the ambient temperature is greater than 77°F and maximum relative humidity is higher than 90 percent during any three days in seven.

Downy mildew, powdery mildew, and blackrot of grapes are also closely related to several atmospheric conditions, making it possible to predict the onset of these diseases. The downy mildew (*Plasmopara viticola*) zoospores that spread the disease are produced only under high humidity (92-100 leaf wetness) and at least four hours of darkness. The optimal temperature for release is between 72° and 75°F. Powdery mildew (*Uncinula necator*) fungal growth occurs at temperatures 43-90°F, but the optimum is between 68 and 81°F and is severely limited if the temperature is greater than 96° F. Black rot (*Guignardia bidwellii*) ascospores are formed in the spring soon after bud break and are ejected after as little as 0.01 inch of rain.

The relationship between diseases and environmental conditions is the basis of the Metos programs for turf and grapes. The heart of these programs is a weather station and data logger. The standard weather station measures air temperature, relative humidity, rainfall, leaf wetness, and day length, and options are available for measurement of leaf and soil temperature, speed and direction of wind, soil moisture, and solar radiation. In addition to helping to forecast disease, such information can provide a basis for predicting insect egg laying, the effectiveness of several fungicides, and the efficiency of spray applications. The weather station, data logger, and warning systems are manufactured by Gottfried Pessl, Schlacht hausgasse 23, 8160 Weiz, Austria and are distributed in the U.S. by Pest Management Supply, Inc., 311 River Drive, Hadley, MA 01035.

Devices which are capable of monitoring environmental conditions 24 hours a day in combination with programmed software are capable of forecasting the potential risk of several diseases and insect attacks. Some of the suitable systems are available from Neogen Corporation, 620 Lesher Place, Lansing MI 48912-1509; Gottfried Pessel; and Automata, Inc., 16216 Brooks Road, Grass Valley, CA 95945-8816.

Other available systems include: (1) MARYBLYT for forecasting apple blight; and (2) Ventem for apple scab prediction. The Pessl instrument, MARYBLYT, and Ventem are distributed in the U.S. by Pest Management Supply Co. MARYBLYT is also distributed by Automata, Inc.

Often the spray program is fashioned so as to prevent damage from diseases which were common in the past. Modifications need to be made in such programs as diseases become apparent, protecting new foliage as soon as disease appears and using materials that may be effective as eradicants. Modifying existing spray programs or starting new ones in such cases will depend on the proper identification of the disease.

VIRUSES, VIROIDS, AND MYCOPLASMAS

Viruses

Viruses are ultramicroscopic particles, consisting of protein shells surrounding ribonucleic acid (RNA), or sometimes deoxyribonucleic acid (DNA). They cannot multiply outside a living cell, but once inside, the protein coat is shed and the core of nucleic acid acts as a template activating the cell to produce virus RNA or DNA and virus coat protein. The nucleic acids and protein coats are quickly combined and the new virus particles repeat the process. In short order, plant cells can be taken over so that they function for the benefit of the virus. The process can seriously affect the function of the plant.

Viral Diseases and Symptoms

Some viruses can be identified by the symptoms produced on the infected plant. Common viral symptoms include reduced growth or stunting, changes in a plant's color, changes in leaf and stem shapes, and death of tissue (necrosis).

Common color changes in leaves are (1) mosaics (yellow-green colors in place of the normal green), (2) mottles or faint mosaics, and (3) stripes or streaks in cereal and grass leaves.

Other color changes are flower mosaics, vein clearing, vein banding, ring spots, and fruit russetting. *Flower mosaics* yield breaks in the coloration of the petals. *Vein clearing* refers to a yellowing, or chlorosis, of the small leaf veins, giving them a netlike appearance. *Vein banding* is a condition wherein the veins remain green while the spaces between them are yellow or chlorotic. *Ring spots* describe a viral symptom that takes the form of spots. Other viruses may cause patterns on leaves or fruits resembling line patterns. *Russetting* of fruits is typified by brown or rust-colored rings.

Abnormal growth or malformation may be apparent on leaves, stems, fruits, and roots. *Raspleaf of cherry* distorts the leaves; *green crinkle virus* distorts apple fruits; and *spotted wilt virus* deforms tomato fruits. *Measles* or *rough bark* deform the twigs and branches of pears.

Necrosis may involve the death of some tissue or the entire plant, as is the case with *Tristeza* of citrus.

Some of the more important viral diseases and their symptoms which can aid in a tentative identification are listed below.

Barley yellow dwarf virus (BYDV) affects barley, oat, pasture grasses, rye, and wheat in various parts of the world, often causing serious losses. It is marked by yellowing and dwarfing of leaves, some of which may turn red or bronze (oat) or yellow (barley, rye, and wheat), while the veins remain green. Other symptoms include a reduced root system, plant stunting, reduced tillering, and sterility or shrunken kernels.

The amount of damage is related to the time of infection in respect to the age of the plant. At an early stage, infection can lead to death of the plant or the complete loss of grain. Later infections may cause reduced yields. The extent of loss is dependent on the amount of damage caused by leaf-spotting fungi, root rotting and winter-killing, which often follow BYDV infestation.

Cucumber mosaic virus (CMV) affects a wide variety of cultivated plants and weeds. It has been reported to affect banana, beans, celery, cantaloupe and other melons, carrot, crucifers, cucumber, eggplant, gladiolus, lettuce, lily, onion, pepper, pumpkin, snap bean, petunia, summer and winter squash, and zinnia. The amount of damage varies with the host, number of vectors, and age of the plant at the time of infection. Infection transmitted at an early age by large numbers of aphids can result in complete loss of susceptible crops, such as cantaloupe. Infection occurring after fruit is set creates minimal loss.

Symptoms of CMV are mottling, distortion, and wrinkling of young leaves. Edges of leaves tend to curl downward. Growth is reduced and the resulting plants are stunted. Tomato leaves become long and stringy. Older leaves of many different plants can become chlorotic and die. There is a tendency to produce few flowers and the resulting fruits may be distorted and misshapen. Those of cantaloupe, cucumber, and squash may also have irregular color markings.

Dasheen mosaic virus (DM) is a serious pest of foliage plants and a few plants of the Araceae family. Symptoms, which vary with the host, include a mild mosaic pattern of light and dark areas, serious distortion of the plant, ring spots, and even death.

Maize chlorotic virus (MCDV), also known as corn stunt, affects all types of corn (field, pop, and sweet). It can shorten plants or cause severe stunting. Leaves become yellow, turning red as the plant matures. Early infection of susceptible corn varieties results in only stunted ears.

The host range of *maize dwarf mosaic virus (MDMV)* is similar to MCDV and symptoms may overlap. Symptoms of MDMV differ from those of MCDV in that the former produces a mottling or mosaic pattern, particularly on young leaves. Leaf discoloration increases as the plant matures, becoming yellowish-green. Plants may tiller excessively with multiple ear shoots bearing few seeds. Plants that are infected early are susceptible to root and stalk rots, leading to the death of the plant.

The hosts of *potato virus Y (PVY)* are pepper, tobacco, tomato, and Irish potato. A common symptom is vein banding. The disease is difficult to identify in the field because of great variation in symptoms brought about by different strains. In combination with *potato virus X*, a symptom results known as rugose mosaic.

The *tobacco etch virus (TEV)* is present in all tobacco growing areas. Hosts, besides tobacco, include pepper, Irish potato, tomato, and weeds such as black nightshade, jimsonweed, groundcherry species, sicklepod, and blue toadflax. Symptoms vary depending upon the strain of virus. Relatively mild symptoms include slight mosaic or etching. More serious ones are severe chlorosis and tattered leaves, which may be followed by death of the plant.

Tobacco mosaic virus (TMV) can infect eggplant, pepper, and tomato, as well as tobacco. The virus also affects a number of weeds–horsenettle, groundcherry species; and black nightshade. The symptoms vary depending on the host plant, strain of virus, or time of infection, and on such factors as plant nutrition, temperature, and light conditions prevailing after infection takes place. Symptoms range from very mild and barely visible to those seriously affecting leaves. The latter may result in reduced yields. Mottling or the typical mosaic of young leaves is the common symptom with tobacco, but with tomato the mottling may also be present on the older leaves. Tomato leaflets also elongate and become very pointed. Fruit set in tomatoes is greatly reduced and often is of poor size and quality. Fruit pulp may turn black.

Tobacco vein mottle virus (TVMV), a disease of tobacco can over-winter in a number of perennial weed hosts. The symptoms vary with the time of infection and the cultivars of tobacco. They are similar to those of vein banding, with which they are often confused. The two can only be differentiated by serological examination.

The hosts of *watermelon mosaic virus (WMV)* are cantaloupe, cucumber, English pea, pumpkin, and summer and winter squash, as well as watermelon. The virus is also present in alyce clover, hairy indigo, and such weeds as balsam pear, creeping cucumber, and showy crotolaria.

Viroids

Several plant diseases are now considered to be caused by viroids, which are much smaller than viruses. They differ from viruses in that they do not have a protein covering. Some of the diseases now thought to be caused by viroids are: chrysanthemum chlorotic mottle, chrysanthemum stunt, citrus exocortis, cucumber pale fruit, potato spindle tuber, and possibly the cadang-cadang disease of coconut palms.

Identification from Symptoms

Tentative diagnosis can be made on the basis of symptoms, which include stunting or dwarfing, distortion of leaves (which may be twisted or rolled), and color changes that involve yellowing and/or chlorosis.

Mycoplasmas

These organisms are midway between bacteria and viruses in size and other properties. They lack cell walls, but they have energy and enzyme systems and are capable of reproducing by forming chains, by budding, or by fission. Injected by aphids and leafhoppers into the phloem of a susceptible host, mycoplasmas multiply rapidly, interfering with the translocation of foods.

Mycoplasmas or mycoplasma-like organisms, also known as phytoplasmas, are now considered to be the cause of lethal yellowing of palms, aster yellows, cherry buckskin, stubborn disease of citrus, corn stunt, mulberry dwarf disease, western-X of peaches, pear decline, big bud of tobacco and tomato, and some witches'-broom.

Mycoplasma Symptoms

Symptoms of mycoplasma diseases, which can form a basis for tentative identification, include reduction in overall growth, loss of green color producing a chlorotic or yellow appearance, and distortions and abnormal growth patterns. Some of these distortions consist of leaf twisting or curling (pear decline), twisting of stems and petioles (aster yellows), proliferation of lateral buds (witches'-broom), and excessive branching of stems and leaves (big bud of tomato).

Identification of Viruses, Viroids, and Mycoplasmas

Often a more positive identification is needed than is possible by comparing symptoms of these diseases. More positive identification is made primarily by using the three basic techniques of indexing, serology, and enzyme-linked immunosorbent assay (ELISA). Indexing basically involves the inoculation of a series of host plants with material from the diseased plant and noting the symptoms produced. Serology utilizes antibodies produced by injecting pure virus into animals. Serum containing antibodies produced in animals is mixed with sap from suspected plants. If the sap contains the same virus producing the antibodies, a precipitate will form, whereas no precipitate is formed if no virus is present or the virus in the sap is different from the one producing the antibodies. Immunosorbent assay first produces a double-antibody sandwich, trapping the pathogen of the test sample in a solid-phase-specific antibody. The trapped antigen is reacted with an enzyme-labeled specific antibody, the amount reacting forming the basis of identification as measured by color changes.

Most methods of positive identification require considerable expertise. In the case of indexing, considerable greenhouse space and longtime care of a number of plants are also needed, putting this method beyond the scope of most growers.

The immunosorbent method offers possibilities of on-location testing, as it is possible to prepare a kit that will provide long-lasting enzyme-specific antibodies and a means of trapping the pathogen.

At present, positive identification for most growers will require collection of suitable samples, packing them carefully, and either quickly bringing the samples or sending them by express means to qualified laboratories. If directions for collection and shipment of samples are not provided by the laboratory, utilize directions as offered in the section on fungi in this chapter. Handling and shipping directions for detection of viruses, viroids, and mycoplasmas in ornamental plants can be obtained from Agdia Inc., Elkhart, IN 46514, who can provide a 24-hour detection service for most organisms.

NEMATODES

Nematodes are round or eel-like worms. Those that parasitize a number of crop plants are microscopic in size. They attack roots, stems, bulbs, foliage, flowers, and seeds, causing a variety of symptoms. The common name usually reflects the type of injury. The *root knot* nematode produces knots or galls on a number of susceptible plant roots. *Stubby root* organisms reduce the root system to short stubby roots. *Stunt* organisms stunt the roots and at times the entire plant. *Lesion* nematodes cause root lesions. *Bulb and stem* nematodes produce stem swellings and shortened internodes. *Bud and leaf* nematodes kill bud and leaf tissue, often distorting the plant.

Damage

The damage to the roots which may be reflected as galls, lesions, excessive branching, damaged root tips, or stunted roots may be visible above ground as excessive wilting, slow decline, foliage yellowing or chlorosis, and poor numbers and/or smaller leaves. In addition to symptoms induced by direct damage, several root rots caused by secondary invasion of bacteria and fungi may also be present.

Damage is most severe in regions with mild winters. Low incidence of freezing and thawing allows nematode numbers to increase. Low soil OM content may also encourage increased numbers, as several organisms that prey on nematodes are favored by OM.

Tentative Diagnosis

While damage caused by root knot organisms are relatively easy to diagnose in the field, that caused by other types may be confused with symptoms caused by infectious disease, nutritional problems, or insufficient oxygen due to compaction or flooding. Positive identification can be made by testing a soil sample for these organisms. Such examination not only will tell whether nematodes are the cause of the problem, but the kinds and numbers present. Such information determines the need for treatment.

Sampling Soils for Nematode Evaluation

Rather than waiting for problems to present themselves, many growers find it useful–at least in certain areas–to routinely test soils for the presence of pathogenic nematodes. Routine soil nematode counts of samples collected prior to planting alert the grower of the need to apply nematicides.

Routine samples and samples collected to determine the cause of damage can be run by many state extension services and commercial laboratories. Directions for sampling and shipping can be obtained from the laboratory in question, or the directions given below can be followed.

1. For routine evaluation, remove 10-20 cores of soil for each unit area by means of a sampling tube, auger, trowel, or spade. The top inch of soil is removed before adding these cores to a clean bucket. Each unit area should comprise no more than 10 acres and be rather uniform in terms of past cropping history, soil type, and fertilizer or manure practices.
2. If samples are being collected to evaluate the cause of a problem, take probes from the root areas of damaged (but not dead) plants. Taking another sample from normal areas containing normal plant roots often helps in evaluating the problem. Including the roots of both troubled and normal-looking plants in their respective samples helps to identify the causes of damage.

   If samples are collected when no plants are present, take random probes of soil from areas where growth of the previous crop was abnormal.

3. Collect samples when soil is neither overly wet nor dry.
4. Collect samples of preplant areas when soil is well prepared and fresh OM has largely decomposed.
5. Adjust the depth of the probes to conform to the crop and type of nematodes. For most crops, use a sampling depth of 8 inches. Take probes to a depth of 3-4 inches when sampling turf. Use depths greater than 12 inches to evaluate the presence of the citrus burrowing nematodes.
6. Mix cores thoroughly before selecting a pint of soil for the composite sample. Include fine roots from live plants in the composite.
7. Place the mixed composite sample in a plastic bag and close it securely. If a Ziplock bag is used, tape the bag tightly to avoid spillage. Do not add moisture to the sample before closing.
8. Label the bag on the outside with a name and an identifying number. Use a black felt-tip marker directly on the bag or a permanent pen or pencil on masking tape stuck to the bag.
9. In addition to the markings on the bag, include information as to crops grown in the past couple of years, a description of existing plants (kind, cultivar, and symptoms); and if treated for nematodes–the kind of treatment and its mode of application. Place all such information in an envelope within the carton.
10. It is important that the sample reaches the laboratory as soon as possible without exposure to heat. Avoid leaving samples in the sunlight or transporting them in a hot trunk. Bring them to the laboratory as soon as possible or ship them by fast mail or express.

WEEDS

A weed has been defined as any plant out of place. Of the 300,000 species of plants inhabiting the earth, about 10 percent possibly exist as weeds. Of these, only about 1800 species cause serious loss of crops. The number that plague individual crop plants usually can be reduced to less than 50, with perhaps ten to 20 that must be controlled in any given area lest serious crop losses ensue.

Weed control is dependent upon proper identification and assessment of numbers. The first step in identification is to classify the type of weed, as certain treatments effective against one type of weed may be valueless against another.

Classification

Weeds can be classified as broadleafs, grasses, or sedges. The broadleafs belong to several families of dicotyledenous plants that have wide leaves, with net-like veins. Many have showy flowers. The grasses have leaves which are much longer than they are wide and the veins are parallel. The sedges are grass-like plants that have solid triangular stems.

Weeds are commonly classified according to the length of their life cycle as well. Annuals have a life cycle of one year. Common annual grasses are barnyard grass, crabgrass, downy brome, foxtail, shattercane, Texas panicum, wild oats, and wild proso millet. Common broadleaf annuals are chickweed, cocklebur, pigweed, lambsquarters, kochia, tansy mustard, velvetleaf, Russian thistle, sicklepod, and ragweed. Biennials need two years to complete their development. Common burdock is a typical biennial broadleaf weed. Perennials live for more than two years. Johnsongrass and quackgrass are perennial grasses. Beggarweed, bindweed, Canada thistle, some morning glories, and nightshade species are perennial broadleaf plants. Purple and yellow nutsedges are perennial sedges.

Common Weeds

Some of the more common weeds are: beggarweed, Bermuda grass, bindweed, Canada thistle, crabgrass, downy brome, hoary cress, foxtail, Johnsongrass, kochia, lambsquarters, leafy spurge, morning glory, pigweed, quackgrass, ragweed, velvetweed, wild onion, wild oats, wild proso millet, and nutgrass.

Parasitic Weeds

Important weeds that parasitize plants are broomrape, dodder, dwarf mistletoe, leafy mistletoe, and witchweed.

Broomrape is a plant without leaves that has clumps of brown, yellow, purple, or white stems that may be 6-20 inches tall. It attacks tobacco, lettuce, tomatoes, and other vegetables by attaching itself to the roots of the host, from which it gets its nourishment while debilitating the host. The mature plant produces many dust-like seeds that can be spread by soil, wind, and water. The seeds can remain dormant for years in the soil until such time that they are stimulated by the host's root excretions.

Dodder looks like a tangle of yellow threads on the upper parts of the host plant. Sometimes dodder will appear almost white and in some instances the threads will have small areas of purple or red. The plant can be a problem anywhere in the temperate or tropical zones. Dodder affects many different kinds of crops, but it is particularly troublesome for alfalfa and clover.

The *mistletoes* affect woody plants, appearing as branched shoots or tufts among the host's twigs. The leafy types infest oaks and other hardwoods, as well as cedar, citrus, cypress, juniper, and pecan. They do not survive cold winters very well, so most that grow in the U.S. are found in the South.

The dwarf type affects conifers and creates considerable losses of pines in the western U.S. The affected trees develop swollen limbs and branch excessively, forming witches'-broom. Infected seedlings may be killed. Other trees may be deformed, stunted, or may break off at the trunk cankers.

Witchweed is a serious pest on corn, sorghum, and sugarcane. It is a colorful plant that grows to a height of about 20 inches, producing bright red flowers. The mature plant produces many small, brown seeds that are readily spread by wind and water. Seeds germinate when stimulated by exudates from hosts' roots. They grow and attach themselves to the hosts' roots. The roots of the witchweed draw sap from the host, weakening it. About 30 days after germination, the weed appears aboveground, at which time it will have produced rootlets that have attached themselves to other roots of the host. The parasitization will stunt the host and make it appear yellowish in color, as though suffering from drought. Seriously affected plants may die, especially in dry weather.

Identification of Weeds

Identification can be made by comparisons with photos, line drawings, and illustrations from a wide selection of materials. State extension services and manufacturers of herbicides provide a number of pamphlets suitable for identification. Several books helpful in diagnosing weeds are listed under Additional Readings.

Control by Limiting the Introduction of Weeds

Weeds may be introduced or increased with crop seeds. Weed counts are often available for new lots of seeds, and growers should shy away from seed lots that may introduce new or additional weeds.

Composts and soils brought into the farm often are a source of contamination. Weed seeds in these materials can be destroyed by heating. A temperature of 175°F maintained for at least 30 minutes will destroy most weed seeds, but some resistant types will have to be heated to 212°F for this period before they are destroyed. Maintaining the materials for several weeks with good moisture and temperatures suitable for growing crops before applying heat simplifies control of resistant seeds.

The various diagnostic procedures which aid in maximizing results from applied herbicides are covered in the next chapter.

Chapter 11

Application of Pesticides

The control of pests is expensive, and carries with it the threat of damage to the environment, danger to applicators or other farm workers coming in contact with chemicals, and possible carryover from certain treatments on foods. Substitution of biological treatments and less-dangerous chemicals for some very toxic chemical treatments, coupled with great care in timing of applications has managed to provide both an abundant and healthy food supply. Nevertheless, limiting pesticide controls to minimum necessary levels is highly desirable. Several diagnostic procedures can be helpful in limiting treatment to those situations which offer high returns to treatments and aiding in applying the correct amount of material.

THRESHOLD DAMAGE

Limiting treatment to those occasions when pest numbers are great enough to do economic damage to the crop can reduce costs and the amount of pesticides applied. The economic threshold, or number of pests necessary to cause economic damage, will vary with the crop, stage of development, the pest, geographic location, weather conditions, and such factors as the time required to carry out the treatment and the speed with which it becomes effective.

Threshold Numbers

Spray treatments to counter insect attack need to be started when numbers threaten economic damage. The devices for continuously measuring environmental conditions discussed in previous chapters can also be programmed to predict the arrival of several insect pests. But the time for initiating controls is best determined by routine scouting by trained personnel, who count the numbers of pests and note the presence of parasites and insect predators as well as the amount of damage inflicted.

Many crops can withstand some loss of vegetation without affecting the value of the crop. Potatoes, for example, can withstand some feeding (10-20 percent of the leaf surface) by the Colorado potato beetle with no decrease in the value of the harvested crop. On the other hand, if the damage is inflicted directly on the harvested part of the plant, very small numbers of pests can cause serious losses. Also, very small numbers of certain pests, such as aphids, leafhoppers, and striped cucumber beetles, can cause serious losses on certain crops because they inject toxins or transmit diseases as they feed. With the latter pest group, special treatments may be necessary as soon as the pest is sighted to avoid exceeding economic thresholds. Treatment should also be started earlier if it is merely preventive in nature or if the pesticide is very slow reacting.

Approximate numbers of nematodes that can cause damage to various crops are given in Table 11.1; numbers for some insects and mites on a few crops are listed in Table 11.2. Since rapidity of development of a number of pests is dependent on climate and the presence of predators and parasites, it is highly desirable to use threshold limits established by local extension or IPM services as a final guidelines for treatment.

EFFECTIVENESS OF PESTICIDES AND pH OF SPRAY SOLUTION

Several pesticides lose their effectiveness very quickly when dispersed in high pH water. The half-life of several pesticides is given in Table 11.3.

The pH of the water can be determined and lowered with a specific amount of acid determined by a titration, or it can be adjusted by the addition of buffers. A suitable pH range is 5.5 to 6.0. Routine addition of buffers assures satisfactory pH with little chance of degradation.

TABLE 11.1. Levels of nematodes that may justify nematicide applications to soils

Kind of nematode	Nematodes per 100 cc of soil	
	Turf	General
Awl	10	
Cyst	any	any
Dagger		100
Lance	40	100
Lesion	80-100	80-100*
Ring	500**	50
Root-knot	80	50
Sheath	80	
Spiral	250	
Sting	10	any
Stubby root	40	100

* Any, if for replant.
** Only 150 for centipede grass
Source: Compilation of data from Dunn, R. A. 1988. Turf nematode management. In *Florida Nematode Control Guide*, IFAS Florida Cooperative Extension Service, Gainesville, FL 32611; A & L Agricultural Laboratories, Memphis, TN and the author's observations.

APPLICATION AT PRESCRIBED RATES

It is important that the pesticide be applied at prescribed rates. Insufficient amounts may fail to give control. Besides wasting precious materials, excessive application may lead to pollution, phytotoxicity, and/or pest resistance.

Recommended rates of application appear on the label. They may be given in rate of active ingredient or as formulated rate per area. Most rates are given as pounds per acre (lbs/a); the active ingredient as pounds active ingredient per acre (lb ai/a). Conversion from active ingredient to formulate material or vice versa is made possible by use of the following equations:

a. For dry formulations,

(1) $\text{lb formulated material/a} = \text{ai/a} \dfrac{100}{\text{ai in the formulation}}$

TABLE 11.2. Threshold values of several insects and mites above which economic loss can occur

Crop	Stage of Development	Insect	Threshold
Cotton*	After 1st bloom	Bollworms or Tobacco budworms	4-20% plants or 5-25% squares infested
	Pinhead squares	Boll weevil	1-4 weevils per trap
	Squares	Boll weevil	10% of squares damaged
	Bolls	Pink bollworm	5-10% of bolls damaged
	Early to large bolls	Spider mites	50-100% leaves infested
	Seedling stage	Thrips	1-5 per plant
Peanuts**	Season long	<u>Foliage feeders</u> Corn earworm	Average of 4-6 worms per row ft.
		Fall armyworm	
		Velvetbean caterpillar	
		Loopers	
		Beet armyworms	
		Yellow-striped armyworm	
		Cutworms	
	Peg/pod	<u>Pod feeders</u> Lesser corn-stalk borers	30% of sites have plants with fresh damage or insects
		Southern corn rootworms	
		Wireworms	
	1st 4 weeks after planting	Thrips	Treat only if new growth shows "tip burn," severe distortion or silvering

Crop	Stage of Development	Insect	Threshold
	Season long	Leafhoppers	Leaf tips turn yellow on more than 30% of the foliage
	Season long	Spider mites	Active mite infestation
Tomato***	Pre-bloom	Armyworms, fruitworms	1 larva/6 plants
	Post-bloom	Armyworms, fruitworms	1 egg or larva per plant
	0-7 true leaves	Tomato pinworms	0.7 larva/plant
	>7 true leaves		0.7 larva/lower leaf
	0-2 true leaves	Leafminers	0.7 larva/plant
	>2 true leaves		0.7 larva/3 terminal leaflets
	Post-bloom	Thrips	>5 flower
	Post-bloom	Stinkbugs	Presence
	Season long	Aphids	>3-4/plant
	Season long	Loopers	1 larva/6 plants
Cucurbits, peppers, or tomato****	Emergence to crop is 2/3 grown	Aphids Green peach, Melon	6 or more per yellow pan trap per day

Source:
*Luttrell, R. G. 1994. Cotton Pest Management: Part 2, A U.S. Perspective. *Ann. Rev. Entomol.* 39: 527-542. This covers U.S. practices and accounts for wide range in threshold limits.
**Weeks, R. Alabama Coop. Extension Service, Headland, AL 36345. Personal Communication.

The foliage feeders are counted in 3 lineal feet of row by briskly brushing down vines, dislodging worms to ground. Counts are replicated at eight to ten sites per field.

The pod-feeding insects are counted by pulling up a plant at each of the sample sites and inspecting it for peg/pod damage, such as borer holes or webs from lesser corn borers. The soil which supported the plant to a depth of 4 inches is sifted for soil insects.

TABLE 11.2. (continued)

***Schuster D. J. and K. Pohronezy. 1989. Practical Application of Pest Management on Tomatoes in Florida. In *Tomato and Pepper Production in the Tropics.* Green, F. K. (ed.) Asian Research and Development Center, Shanhua, Tainan, Taiwan.

****Simons, J. N., JMS Flower Farms, Vero Beach, FL. Personal communication. Control of aphids is for virus control (Papaya ringspot and Zucchini yellow mosaic for cucurbits; Tobacco etch, Pepper mottle, and Potato virus Y for peppers; and Tobacco etch and Potato Virus Y for tomatoes) in southern Florida. Trap is made from an aluminum pie plate sprayed with yellow paint and containing a 50/50 mixture of water and ethylene glycol. If there are multiple plantings growing simultaneously, scouting is continued until the final crop is two-thirds grown.

b. For liquid formulations,

(2)   gal formulated material/a = $\dfrac{\text{lb ai/a}}{\text{lb ai/gal formulation}}$

Or the amounts can be taken from Table 11.4.

APPLICATION OF PRESCRIBED AMOUNTS

The application of the prescribed amount can be assured by several diagnostic procedures that differ slightly for greenhouse and field applications.

Field Application

The application rate of the formulated material can be carefully monitored by periodic calibration of application equipment (see Calibration of Application Equipment in this chapter). Once the equipment is calibrated, constant ground speed and pressure must be maintained. Spray nozzles must be kept clean and periodically checked to determine whether volume delivered is within 5 percent of the original output. The frequency for checking nozzles depends on the abrasiveness of the materials applied and the type of nozzles used. Equipment must be maintained in good order with proper cleaning after each use to avoid corrosion.

Pesticide applications such as aerosols, fogs, and smokes in greenhouses or other enclosed areas must be adjusted to the volume

TABLE 11.3. The half-life of several pesticides as affected by tank pH

Pesticide	pH	Half-life
Dylox/Proxol	6.0	3.7 days
	8.0	1 hour, 3 minutes
Imidan	4.5	13 days
	8.3	4 hours
Lorsban/Dursban	7.0	35 days
	8.0	1.5 days
Sevin	6.0	100-150 days
	9.0	24 hours
Captan	4.0	4 hours
	10.0	2 minutes
Orthene	3.0	66 days
	9.0	3 days

Source: Dill, R. A. 1990. Spray problems may be pH. *Farm Chemicals* (May) 153. Reprinted by permission.

of the structure in addition to the conditions cited for field application. The volume of most greenhouses can be determined by multiplying the floor area (length × width) by the average height (height of the sidewall + the height of the ridge divided by two). For example, the volume of a greenhouse that has floor dimensions of 40 feet by 60 feet and has a sidewall 8-feet high and a ridge of 15 feet, will be 27,600 cubic feet, obtained as follows:

$$40 \times 60 = 2400 \text{ square feet of floor space}$$
$$\frac{8 + 15}{2} = 11.5 \text{ feet average height}$$
$$2400 \times 11.5 = 27,600 \text{ cubic feet.}$$

CALIBRATION OF APPLICATION EQUIPMENT

Calibration of equipment will help to avoid the unnecessary application of a pesticide. Calibration helps ensure uniform coverage and helps provide desirable rates. Suggested methods of calibrating the different application systems are briefly outlined below.

TABLE 11. 4. Amounts of formulated materials required to supply varying amounts of active ingredients on an acre basis

A. DRY MATERIALS*

Pounds active ingredients (ai/a)

Formulation	1		2		4		5		10	
	lbs	oz	lbs	oz	lbs	oz	lbs	oz	lbs	oz
2 G	50		100		200		250		500	
10 G	10		20		40		50		100	
15 G	6	11	13	5	26	11	33	5	66	11
35 SP, WP, or WDG	2	14	5	11	11	7	14	5	18	9
40	2	8	5		10		12	8	25	
50	2		4		8		10		20	
65	1	9	3	1	6	2	7	11	15	6
75	1	5	2	11	5	5	6	11	13	5
80	1	4	2	8	5		6	4	12	8
90	1	2	2	4	4	7	5	9	11	2

B. LIQUID MATERIALS

Formulation	1		2		4		5		10	
	qts	oz	qts	oz	qts	oz	qts	oz	qts	oz
EC, F, MC, or SC										
1.5 (16%)	2	21	5	11	10	21	13	11	26	21
1.8 (24%)	2	7	4	14	8	28	11	4	22	7

Formulation										
2 (25%)	2		4	21	8	11	10	21	20	11
3 (33%)	1	11	2		5		6		13	
4 (41%)	1		2	19	4		4	1	8	2
5 (55%)		26	1		3	6	4		8	
6 (62%)		21	1	11	2	21	3	11	6	21

* Dry materials; G = granular; SP = soluble powder; WP = wettable powder; WDG = water-dispersible granules.
**Liquid materials; EC = emulsifiable concentrate; F = flowable concentrate; MC = miscible concentrate; SC = soluble concentrate. Amounts are in fluid measurements, where 16 ounces = 1 pint; 2 pints = 1 quart; 4 quarts = 1 gallon.

Source: Stamps, R. H. 1989. Pesticide application rate conversions. *Florida Grower and Rancher*, (February) 82. Reprinted by permission.

Boom Sprayers

Before attempting to calibrate a spray rig, it must be checked to see that it is operating satisfactorily. Prerequisites for satisfactory operation are a clean tank, screens, boom, and nozzles; hoses free of leaks; and a pump able to maintain the desired pressures. The nozzles are checked for uniform delivery by placing containers under each nozzle, catching the water emerging in a ten- to 20-second period while the applicator engine is running and pressure is at the desired rate. Nozzles are replaced if delivery is more than 5 percent of rated volume delivery. The calibration is accomplished in a field run over a premeasured course by measuring or weighing the contents of the sprayer before and after the test. The nozzle pressure and speed of travel must be uniformly maintained during the test. The pressure selected depends on the size of the spray tips and gpm needed to apply the allotted amount per acre. The detailed procedure using a 300-foot course is presented in Figure 11.1. Measuring contents by a mark will only be correct if the applicator is parked in the same level spot used at the beginning of the test. Weighing the contents rather than measuring them is a more accurate means of calibration in most cases.

Distances other than 300 feet can be used for calibration. The area covered is determined by the equation:

$$(3) \quad \text{Area (acres)} = \frac{\text{distance traveled (feet)} \times \text{swath width (feet)}}{43,560}$$

If 10 gallons were applied in a test over a 500-foot course and the swath width was 20 feet, then 0.23 acres were covered and 43.47 gallons of water were applied per acre.

$$(4) \quad \frac{500 \times 20}{43,560} = 0.23 \text{ acres} \qquad \begin{array}{l} 10 : 0.23 \text{ as X:1} \\ X = 43.5 \text{ gallons per acre} \end{array}$$

The gallons per acre can also be calculated by multiplying the gallons used in the test strip by 43,560 and dividing the answer by the area sprayed in square feet. In the above trial:

FIGURE 11.1. Calibration of a sprayer based upon gallons needed to refill spray tank after a 300-foot strip has been sprayed

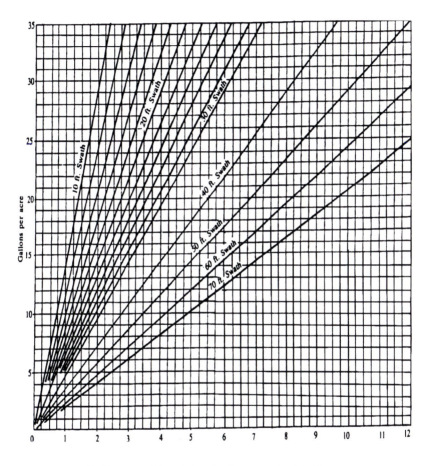

Gallons used to refill tank after spraying 300 feet.

Source: Agricultural Engineering Department. A. U. Agricultural Expt. Sta. and Reproduced by Courtesy of Auburn University Agric. Extension Service.

FIGURE 11.1. (continued)

INSTRUCTIONS FOR CALIBRATING A BROADCAST SPRAYER
(Boom or Boomless)

1. Measure out a 300-ft. distance in the field to be sprayed or in an area with a surface similar to that in the field to be sprayed.
2. With water in sprayer and throttle set at operating speed, operate sprayer and adjust pressure to the desired setting (10 to 60 lbs.).
3. Fill tank completely full or to a mark on a gauge stick.
4. Put tractor in operating gear with throttle set at operating speed and operate the sprayer over the 300-ft. distance.
5. Measure the amount of water to the nearest quart used to refill the tank or to fill to the point on the gauge stick.
6. Then read chart according to the following instructions:
    a. Move across the bottom scale to the gallons used to refill tank.
    b. Then move up to the line with your swath width.
    c. Then move across to the vertical scale and read gallons per acre.

Example: If you used 3 1/2 gallons to refill tank and if you are using a 30 ft. swath, the chart shows you are applying 17 gallons per acre.

$$10 \times 43,560 = 435,600$$
$$500 \times 20 = 10,000$$
$$435,600 \div 10,000 = 43.56 \text{ gallons per acre}$$

The calculations can be avoided if a specific test length is selected based on nozzle spacing as per Table 11.5. The time needed to cover the length selected must be accurately measured with a stopwatch. Instead of weighing or measuring the contents remaining in the tank, obtain the amount of liquid in ounces dispensed from one nozzle while the engine rpm is the same as used in the field and the length of time is similar to that needed to cover the course. The rate per acre in gallons is equal to the ounces of water collected per nozzle.

Airblast Sprayers

Airblast sprayers are calibrated in very much the same manner as the boom sprayers in that a test area is covered with a full tank and the quantity of material to refill the tank is weighed or measured. In

TABLE 11.5. Length of test course needed for different nozzle spacings so that ounces delivered per nozzle equals gallons per acre*

Nozzle spacing	Test course length
in	ft
10	408
12	340
14	291
16	255
18	227
20	204
22	185
24	170
26	157
28	146
30	136
32	127
34	120
36	113
38	107
40	102

* If liquid is collected from nozzle for same length of time.

operating the airblast sprayer, special attention must be paid to the following:

1. Operate the sprayer at wind speeds less than 5 miles per hour.
2. Be sure that coverage by the machine equals the intended swath. If in doubt as to the extent of coverage, use black plastic indicator sheets in the rows at various distances from the machine course to determine the deposit effectiveness.
3. With vanes and nozzles set to give the best coverage, maintain the correct pressure to give the proper droplet size for uniform coverage of the test swath.
4. For tall row crops, set the course so that material is blown down the rows instead of across them.

Granular Applicators

Broadcast application of granular materials can be determined by the following:

1. Using the application rate from the label and data from the application operator's manual, set the dial or feed gate to apply the desired rate.
2. Fill the hopper to a given level and mark it while the equipment is parked on a level surface.
3. Apply material to an accurately measured length of marked test area, operating at the same speed that the applicator will normally travel.
4. Carefully weigh the material needed to refill the hopper to the previously marked level.
5. Determine the application rate with the equations given below.

(5) Acres of area covered $= \dfrac{\text{Length (feet)} \times \text{width (feet) of application}}{43,560}$

(6) Application rate $= \dfrac{\text{Pounds of material needed to refill hopper}}{\text{Area covered in acres}}$

The acreage for a test covering a 300-foot by 20-foot swath* = 0.138, derived by multiplying 20 by 300 and dividing by 43,560. If 10.5 pounds of material is needed to refill the hopper after it has traveled the test course, then the rate per acre equals 76 pounds, obtained by dividing 10.5 by 0.138.

If the application rate as determined above does not equal the desired rate, change the gate valve or opening and repeat the test.

Row or band application of granular materials can be calibrated as follows:

1. As for broadcast applicators, determine the application rate and set the dial or feed gate according to the label and operator's manual.
2. Fill the hopper half full and start the application until all units are feeding.

---

*For gravity or drop spreaders, the width of the swath is the same as the width of the spreader. The width of the swath for spinner applicators equals the working distance or distance between runs.

3. Stop the application, remove the feed tubes at the hopper, and attach a bag over the hopper openings.
4. Operate the machine over a measured distance, maintaining the speed that will be used for application.
5. Weigh the amount delivered from each hopper. Hoppers should deliver about equal amounts. If not, make adjustments and recalibrate.

$$(7) \quad \text{Area} = \frac{\text{Length of test (feet)} \times \text{band width (feet)} \times \text{band numbers}}{43,560}$$

$$(8) \quad \text{Application rate} = \frac{\text{Amount applied in the bands (total collected)}}{\text{area}}$$

It is difficult to collect the material from power centrifugal broadcast applicators while moving in the field. A suitable collection can be made while the applicator is stationary. First, it is necessary to determine the length of time it takes to move an empty applicator over the test course at the desired speed of operation. Then, the applicator is filled and the granules are collected while the equipment operates in a stationary position for the predetermined length of time that it took to cover the test course. The data are used in the equations above, but in this case the effective width of the swath, rather than the width of the machine, must be entered in the first equation.

Changing over from broadcast to band applications requires the following equation:

$$(9) \quad \frac{\text{Band width (inches)}}{\text{Row spacing (inches)}} \times \begin{array}{c}\text{broadcast}\\\text{rate}\\\text{per acre}\end{array} = \begin{array}{c}\text{amount}\\\text{needed}\\\text{per acre}\end{array}$$

For aerial application the following equations will be useful:

$$(10) \quad \text{Acres covered} = \frac{\text{length of swath (miles)} \times \text{gal/a}}{8.25}$$

$$(11)\ \text{Acres/min} = \frac{2 \times \text{swath width} \times \text{mph}}{1000}$$

$$(12)\ \text{gpm} = \frac{2 \times \text{swath width} \times \text{mph} \times \text{gal/a}}{1000}$$

COVERAGE

The effectiveness of many pesticides is closely related to the degree of coverage of the intended target. Control of a great number of pests is possible only if the spray effectively covers the undersides of leaves, where damage is incurred. Also, control of fungi and a number of insects depends on a continuous film of protection.

Coverage can be assessed by using a hand lens (10-20 power) to locate droplets or dust particles soon after a pesticide is applied. Evaluation of spray coverage is greatly enhanced if fluorescein is used as a chemical marker for observing spray droplet distribution.

Evaluating Coverage with Fluorescent Light

The test is best performed at night. The sodium salt of fluorescein is added at the rate of 1 cup (237 cubic centimeters) to every 50 gallons of water in the spray tank and the spray is applied using standard pressures, ground speed, boom height, and nozzles. Examination of both the top and undersides of the leaves should be made within 15 minutes of application. A special hand-held flashlight (Eveready Emergency Light equipped with a Phillips F6T5/BL-B black light bulb) provides the fluorescense. The distribution of the droplets can be examined by eye, although this might limit the observance of very small droplets (50-60 microns) unless these droplets coalesce. Placing leaves under a dissecting microscope eliminates this problem. (Personnel handling fluorescein should use rubber gloves and protective clothing because it is a possible carcinogen.)

The fluorescent light test enables a grower to quickly evaluate the current spraying tactics. Ineffective coverage can be corrected by changing spray nozzles, boom height, pressure, or ground speed.

CARRYOVER OF PESTICIDES

The carryover of pesticides is of concern because of three basic reasons; (1) the amount of herbicide remaining in the soil after a crop is completed can adversely effect the health of the succeeding crop; (2) pesticides remaining on foods can affect the health of consumers; and (3) pesticides filtering into drinking water or remaining in the environment can pose threats to man and wildlife.

Soil Carryover

The amount of pesticide remaining in the soil is primarily a function of persistence and sorption.

Persistence defines the length of time over which a pesticide can persist or fail to be degraded by such factors as sunlight and microbial action. The period of degradation is usually listed as a half-life, or the time needed to completely destroy one half of an applied pesticide. The half-life of a pesticide depends on the nature of the chemical.

Sorption is the process by which pesticides stick to soil particles, primarily to OM and clay. In any given soil, sorption is largely affected by the pesticide's solubility and the amount of water falling or added after application.

The amount of pesticide sorption can be calculated by use of the partition coefficient, expressed as:

$$Kp = (Koc)\,(\%OM)\,(0.0058), \text{ where}$$
$$Kp = \text{index for sorption}$$
$$Koc = \text{partition coefficient}$$
$$\%\,OM = \text{percent organic matter in the soil}$$

If Koc equals 0, pesticides are not absorbed and therefore will leach as any nonabsorbed ion. Pesticides with larger Koc values and a long half-life tend to remain near the ground surface and, therefore, are more subject to runoff losses which can contaminate lakes and streams. Pesticides with small Koc values and a long half-life are readily leached through the soil and contaminate ground waters. The fate of nonpersistent pesticides with a short half-life depends on the occurrence of heavy rainfall or irrigation soon after application. If there is no large supply of water to move the pesticides downward soon after application, these materials remain in the root zone where they are degraded without harming water. The Koc and sorption values of a number of pesticides are presented in Table 11.6, and the relative danger to surface and ground waters of the different types of pesticides is shown in Table 11.7.

Measuring Carryover of Pesticides

To avoid damage to a succeeding crop, areas treated with herbicides that will be planted with a sensitive crop need to be tested for the presence of carryover herbicides. For maximum information, separate composite samples are collected from soils varying widely in OM, moisture, or clay content. Samples are also collected from untreated fields or strips and are used as checks. The evaluation of the soil sample can be made by bioassay, immunoassay, or chromatographic testing.

Bioassay

Simple tests for detecting residues can be carried out on the farm by growers. Samples of soil are placed in containers (plastic cups with a drainhole will do) and filled to within an inch of the top. Seeds of the intended crop are planted in the containers. Carryover is estimated from the appearance of plants as they develop. Carryover also can be estimated by planting a crop susceptible to the herbicide. Crops suitable for evaluating carryover of several herbicides are ryegrass, oats, or soybean for *triazine*; corn, oats, rye, or sorghum for DMA; alfalfa or corn for the herbicides *Classic* and *Scepter*; corn, oats, or sunflower for *Command*. Moisture, temperature, and light must be kept satisfactory for germination and growth, and the emergent plants observed for abnormal symptoms. Symptoms indicating carryover problems are poor germination; poor root development; distorted, chlorotic, or discolored leaves; poor growth; and death of the plant.

The test is simple but requires considerable time. Sample collection needs to be delayed sufficiently to allow for maximum breakdown of the chemical. This delay, along with the time needed to

TABLE 11.6. Persistence and sorption of some pesticides in soils

Common name	Trade names	Koc	Half-life
NONPERSISTENT (half-life 30 days or less)			
dalapon	Basfapon, Dowpon	1	30
dicamba	Banvel	2	14
chloramben	Amiben	15	15
metalaxyl	Ridomil	16	21
aldicarb	Temik	20	30
oxamyl	Vydate	25	4
propham	Ban-Hoe, Chem-Hoe	60	10
2,4,5-T	Dacamine 4T, Trioxone	80	24
captan	Orthocide, Captanex	100	3
fluometuron	Cotoran, Lanex	100	11
alachlor	Alanex	170	15
cyanazine	Bladex	190	14
carbaryl	Sevin	200	10
iprodione	Rovral	1000	14
malathion	Cythion	1800	1
methyl parathion	Penncap-M, Metacide	5100	5
chlorpyrifos	Lorsban, Dursban	6070	30
parathion	Thiophos, Bladan	7161	14
fluvalinate	Mavrik, Sour	100000	30
MODERATELY PERSISTENT (half-life >30 but <100 days)			
picloram	Tordon	16	90
chromium-ethyl	Classic	20	40
carbofuran	Furadan, Curaterr	22	40
bromacil	Hyvar, Bromax	32	60
diphenamid	Enide, Rideon	67	32
ethoprop	Mocap	70	50
febsulfothion	Dasinit	89	33
atrazine	Attrex	100	60
simazine	Princep	138	75
dichlorbenil	Casoron	224	60
linuron	Lorox, Aflon	370	60
ametryne	Evik	388	60
diuron	Karmex	480	90
diazinon	Basudin, Spectracide	500	40
prometryn	Caparol, Primatol Q	500	60
fonofos	Dyfonate	532	45
chlorbromuron	Maloran	996	45
azinphos-methyl	Guthion	1000	40
cacodylic acid	Bolate, Bolls-Eye	1000	50
chlorpropham	Beet-Kleen, Furloe	1150	35
phorate	Thimet	2000	90
ethalfluralin	Solanan	4000	60
chloroxuron	Tenoran, Norex	4343	60
fenvalerate	Extrin, Sumitox	5300	35
esfenvalerate	Asana	5300	35
trifluralin	Treflan	7000	60
glyphosate	Roundup	24000	47

TABLE 11.6 (continued)

PERSISTENT (half-life >100 days)			
Common name	Trade names	Koc	Half-life
fomesafen	Flex	50	180
terbacil	Sinbar	55	120
metsulfuron-methyl	Ally, Escort	61	120
propazine	Milogard, Primatol P	154	135
benomyl	Benlate	190	240
monolinuron	Aresin, Afesin	284	321
prometon	Primatol	300	120
isofenphos	Oftanol	408	150
fluridone	Sonar	450	360
lindane	Isotox	1100	400
cyhexatin	Plictran	1380	180
procymidone	Sumilex	1650	120
chloroneb	Terraneb	1683	180
endosulfan	Thiodan, Endosan	2040	120
ethion	Ethion	8800	350
metolachlor	Bicep	85000	120

Source: Hornsby, A. G. 1990. Managing pesticides for crop production and water quality protection. *Florida Grower & Rancher* (February) 83:19,20,22,24. Reprinted by permission.

TABLE 11.7. The potential of pesticides for contaminating ground and surface waters based on their persistence and sorption

Persistence period	Sorption	Potential Impact	
		Groundwater	Surfacewater
Nonpersistent	Low-moderate	Low	Low
Nonpersistent	Moderate-High	Low	Moderate
Mod. persistent	Moderate-High	Moderate	Moderate
Mod. persistent	Low-Moderate	High	High
Persistent	Moderate-High	Moderate	High
Mod. persistent & persistent	Low-High	Site-specific conditions determine impacts	

Source: Hornsby, Managing Pesticides.

carry out the test, mean that the information may not be available until it is too late to be of benefit.

Immunoassay

Use of an immunoassay greatly shortens the length of time required to detect residues of herbicides and other pesticides. The system relies on antibodies that bind the active ingredients of the different pesticides. Enzyme-linked assays (ELISAs), which have become the preferred method of pesticide immunoassay, use the enzyme as a label to detect the pesticide's presence as it produces a color change. Enzymes compete with pesticides for binding sites on the antibodies and indicate the amount of reactants that form the antibody-herbicide complex. Added chromogens impart a blue color to the sample-antibody-label mixture, with the particular shade of blue indicating the amount of pesticide present.

The soil sample can be sent to a laboratory that conducts immunoassays for herbicides, or a suitable test can be run at the farm with different test kits. A separate test kit is needed for each type of herbicide. Kits for detecting a number of herbicides are now available from Agri-Diagnostics Associates, Cinnaminson, NJ 08057; Millipore Corp., Bedford, MA 01730; Neogen Corp., Lansing, MI 48912; and Enzytec Inc., Kansas City, MO 64110.

Chromatographic Tests

Chromatographic testing also greatly shortens the time needed to detect herbicide residues. The tests require a very expensive analytical instrument and considerable expertise in using it, limiting the running of such tests to laboratories. There are several state extension services and private laboratories capable of running chromatographic tests. The method is highly reliable but costs are about $100 per sample as compared to $10-15 for immunoassay tests.

Pesticides in Food

The waiting period between the last application of a pesticide and the harvest of the crop allows for decomposition of the pesticide so that amounts remaining usually fall within accepted tolerances. The long time needed to test for pesticides using laboratory methods has generally deterred growers from testing for residues. The recent introduction of the immunoassay kits by Agri-Diagnostics and Millipore, which enables growers to quickly determine the presence of various insecticides, fungicides, and herbicides, opens the possibility for the grower to determine whether harvested produce may exceed accepted limits. Such tests can be made prior to harvest and if residues exceed limits, extra waiting periods can be allowed. If this approach is not practical, special trimming or washing procedures to lower residues may be used.

Pesticides in the Environment

Recent regulations, aimed at reducing the pollution of the environment (especially drinking water) by agricultural chemicals, place greater responsibility on growers and applicators of chemicals. Monitoring water and soil for pesticides can help avoid costly fines. The introduction of immunoassay kits makes it possible for the grower to monitor the carryover of pesticides, but most growers will find it useful to hire consulting services for such purposes.

Additional Reading

Anonymous. *Identification and Control. Broadleaf Weeds in Turf.* DowElanco, Indianapolis, IN 46268-1189.

Brooks, A. and A. Halstead. 1980. *Garden Pests and Diseases.* Simon and Schuster, Gulf and Western, Rockefeller Center, 123 Avenue of the Americas, New York, NY 10020.

Carlile, W. R. 1988. *Crop Diseases.* Edward Arnold, a division of Hodder & Stoughton, London, Baltimore, Melbourne, Aukland.

Colvin, D. L. 1989. *Florida Weed Control Guide.* IFAS. Florida Cooperative Extension Service, Gainesville, FL 32611.

Compendium of Crop Diseases. APS Press, St. Paul, MN 55121. A series of monographs on diseases of individual crops, ornamental foliage, and turf. Slide sets to identify diseases of about ten different crops are also available.

Davidson, R. H. and W. F. Lyon. 1987. *Insect Pests of Farm, Garden, and Orchard.* 8th ed. John Wiley & Sons, P.O. Box 6792, Somerset, NJ 08873-9976.

Dunn, R. A. 1988. *Florida Nematode Control Guide.* IFAS. Florida Cooperative Extension Service, Gainesville, FL 32611.

Ennis, W. B. (ed.). 1979. *Introduction to Crop Protection.* Amer. Soc. of Agronomy, Crop Science Soc.of Amer., Madison, WI.

Ferris, C., R. E. Rice, P.H. Twine, J. M. Pozzi, Jr., C. E. Kennett, and H. S. Elmer. 1984. *Insect Identification Handbook.* University of California, Agriculture and Natural Resources, 6701 San Pablo Ave., Oakland, CA 94608-1239.

Flint, M. L. 1990. *Pests of the Garden and Small Farm.* University of California, Agriculture and Natural Resources, Oakland, CA 94608. This book, which emphasizes alternate methods of pest management, has 250 color photographs, 28 black-and-white photographs, and 90 line drawings, many of which will help in diagnosis.

Foster, R., E. Knake, R. H. McMarty, J. S. Mortvedt, and L. Murphy (eds.). 1995. *Weed Control Manual.* Meister Publishing Co., Willoughby, OH 44094.

Grubler, M. D., A. H. McCain, H. D. Ohr, A. O. Paulus, and B. Teviotdale. Revised 1991. *California Plant Disease Handbook and Study Guide for Agricultural Pest Control Advisors*. Cooperative Extension Service, University of California, Division of Agriculture and Natural Resources, Publication No. 4046, Oakland, CA 94608-1239.

Hamlen, R. A., D. E. Scott, and R.W. Henley. *Detection and Identification of Insects and Related Pests of the Commercial Foliage Industry*. IFAS. Florida Cooperative Extension Service. University of Florida, Gainesville, FL 32611.

James, W. C. 1979. *A Manual of Assessment Keys for Plant Diseases*. American Phytopathological Society, St Paul, MN 55121.

Johnson, W. T. 1979. *Insects that Feed on Trees and Shrubs*. Cornell University Press, Ithaca, NY and London.

Kader, A. A. (ed.). 1992. *Postharvest Technology of Horticultural Crops*. University of California, Division of Agriculture and Natural Resources, Publication No. 3311, Oakland CA 94608-1239. This publication has 16 pages of color photographs as well as descriptions and controls of postharvest diseases affecting fruits and vegetables.

Klingman, G. C., F. M. Ashton, and L. J. Noordhoff. 1982. *Weed Science*. John Wiley & Sons, Inc., P.O. Box 6792, Somerset, NJ 08873-9976.

Kranz, J., H. Schmutterer, and W. Koch (eds.). 1977. *Disease, Pests, and Weeds in Tropical Crops*. John Wiley, Chichester, New York, Brisbane, Toronto. An excellent treatise on the damage caused by bacterial and fungal disease, insects and mites, nematodes and mycoplasm disease to tropical and semitropical crops. Includes detailed descriptions and many photographs, some of them in color.

Kucharek, T. A. and G. Simone. 1988. *Florida Plant Disease Control Guide*. IFAS. Florida Cooperative Extension Service, Gainesville, FL 32611.

Lorenz, O. A. and D. N. Maynard. 1988. *Knott's Handbook for Vegetable Growers*. 3rd ed. John Wiley & Sons, Inc., New York, NY.

Lucas, G. B., C. L. Campbell, and L.T. Lucas. 1985. *Introduction to Plant Disease*. Avi Publishing Co. Inc., Westport, CT.

Nelson, P. V. 1978. *Greenhouse Operation and Management.* Reston Publishing Co., Prentice Hall, Reston, VA.

Rice, R. P., Jr. 1986. *Nursery and Landscape Weed Control Guide.* Thompson Publications, P.O. Box 9335, Fresno, CA 93791.

Short, D. E., P. G. Koehler, R. C. Wilkinson, W. Doxon, S. Foltz, C. W. Fatsminger, F. A. Johnson, and R. Mizell. 1988. *Florida Insect Control Guide.* IFAS. Florida Cooperative Extension Service, Gainesville, FL 32611.

Smith, M. D. (ed.). 1983. *Ortho Problem Solver.* Ortho Information Services, Chevron Chemical Co., 575 Market St., San Francisco, CA 94105.

Yepsen, R. M., Jr. (ed.). 1984. *Encyclopedia of Natural Insect and Disease Control.* Rodale Press, Emmaus, PA. A rather complete book of disease and insect activity affecting a great number of crops. Includes descriptions, line drawings, and color photographs.

SECTION V:
DIAGNOSTIC PROCEDURES
FOR EVALUATING
THE PLANT ENVIRONMENT

Crop yields are greatly affected by environmental conditions to which plants are exposed. Light, temperature, soil nutrients, soil air and moisture, air humidity, air oxygen and carbon dioxide, air pollutants, and wind all affect plants in various ways. Plants vary as to their responses to these components of the environment, but all have ideal ranges which need to be maintained if the plant is to realize its full potential as programmed by its genetic makeup.

Various diagnostic procedures are available that can measure the separate components of the environment and thus assist the grower in maintaining ideal conditions. Those that measure soil air and nutrients have been largely covered in Section I and that of soil water in Section III. In this section, we will consider diagnostic procedures that can help the grower avoid problems with light, temperature, humidity, carbon dioxide, air pollutants, and wind.

Chapter 12

Light

Light is the essential energy source for green plants as they convert carbon dioxide to carbohydrates in the process of photosynthesis. Light also plays a part in other important functions. Two of the more important ones are the reduction of nitrate-nitrogen, and the synthesis of vital hormones. Light reduces nitrate-nitrogen (NO_3-N), which cannot be used, to the amino form (NH_2-N), utilized by plants to form the amino acid that is so crucial to the formation of various nitrogenous compounds. Hormones are essential for plants, as they either initiate or control the germination of certain seeds, the form and size of the plant, the flowering and fruiting of many plants, the initiation of dormancy in buds and seeds, and the onset of senescence.

COMPONENTS OF LIGHT

The effects of light noted above are dependent on quality (wavelength), intensity (radiant energy), and duration of light (photoperiod).

Light Quality

Wavelengths, varying in range from 280-1000 nanometers (nm) affect plants in different ways. Although the red end of the spectrum (610-700 nm) is more efficient in photosynthesis, the blue-violet end (400-510 nm) has a greater formative effect on the plant. Eliminating the red end of the spectrum leads to the formation of almost normal appearing plants. But plants grown in daylight from which the blue-violet end of the spectrum below 529 nm is eliminated tend to elongate more rapidly in the first few weeks after germination, and have much thinner stems, poorly differentiated internal tissues, lower fresh and dry weights, and very poor development of flowers, fruits, and storage organs.

Intensity

The intensity or quantity of light varies with the source of light, distance from the source, background affecting the reflection of light, and interfering substances which can filter or obscure the light. Intensity is expressed in terms of foot candles (fc), lux units, or as watts per unit area for electrical energy. An fc is the illuminance of one candle at a distance of 1 foot, or 1 lumen/square foot. A lux is the illuminance of 1 lumen/square meter. (A lumen equals a unit of luminous flux emitted by a standard source of 1/680 light-watt). One lux equals 0.0929 fc and 1 fc equals 10.76 lux. A fc also equals about 0.03 watts per square meter (w/m^2).

The universal source of light is the sun, but much of its light is affected by dust particles and moisture in the air. Even on clear days, the intensity increases from morning to noon and decreases from noon to night. The intensity of the sun, which is about 10,000 fc, is diminished by various structures. The fc values at various locations and of different sources of light are presented in Table 12.1.

Light intensity affects plants in many ways, but the symptoms of insufficient light intensity vary according to whether the plant is a monocot (emerges with one leaf) or a dicot (emerges with two leaves). The dicot symptoms of low light intensity are elongated stems, lengthened leaf petioles, and small, undeveloped, pale leaf blades. The monocot symptoms are narrow, elongated leaves along with normal appearing stems. In both types of plants, roots are poorly developed, and although leaves have an abundance of chlorophyll, flowering and fruit development may fail to occur.

As light increases up to medium intensity, stems become shorter, roots are better developed, leaf and plant heights tend to reach maximum size, and all tissues are better differentiated. At very high intensities, plants tend to be shorter and stockier, with smaller and more compact leaves, thicker stems, and fully developed roots and

TABLE 12.1. Intensity (fc) of different light sources

Light source	Location	fc
Sunlight	outdoors–full sun	10,000
	greenhouse–winter overcast	1000
	home indoors	
	1 ft from north window	200-500
	3 ft from north window	100-180
	1 ft from south window–shade	500-900
	1 ft from east window	250-400
	2 ft from east or west window	150-250
Incandescent bulb		
75 watt	1 ft away	150
100 watt	3 ft away	40
150 watt	3 ft away	90
300 watt	3 ft away	180
Fluorescent bulb		
40 watt 1 tube	1 ft away	120
	2 ft away	75
40 watt 2 tubes	1 ft away	240
	2 ft away	120
40 watt 4 tubes	1 ft away	550
	2 ft away	320
Maximum Photosynthesis		1200

Source: A compilation of data from Kaufman, P. B., T. L. Mellichamp, J. Glimn-Lacy, and J. D. LaCroix. *Practical Botany.* 1985. Reston Publishing Company, Reston, VA; and Janick, J. *Horticultural Science*, Third edition. W. H. Freeman and Co.,San Francisco, CA.

storage organs. Plants also tend to attain maximum flowering and fruiting under conditions of high light intensity.

OPTIMUM LIGHT FOR DIFFERENT PLANTS

Species of plants vary with regard to the intensity of light that is optimum for them. Many plants do best in full sunlight, but a number will do well only with greatly reduced light intensity. In between are a great number of plants that perform best in varying degrees of shade.

The amount of light needed to maintain a plant at a certain stage of growth is much less than is needed to continue its growth. Light intensity requirements for growth or maintenance of foliage plants suitable for interior plantings are given in Table 12.2.

Acclimatizing Plants to Reduced Light

Many plants can be acclimatized to different light levels if the change is made gradually. This approach is being used in the production of plants for interior use. For example, placing a sun-grown *Ficus benjamina* indoors usually will result in substantial leaf loss. Placing the sun-grown plants under 80 percent shade for five weeks before placing them indoors reduces leaf drop by about 50 percent. The acclimatized ficus has both small, thick, light green sun foliage and thin, large, dark green shade foliage. Such plants will survive indoors if provided with 75-100 fc of light for ten to 12 hours each day.

Plants that can survive the poor light intensity associated with most interiors need to be maintained with low fertilization rates, but production of plants with very low light and fertilizer programs is very difficult. A compromise allowing fairly rapid production uses

TABLE 12.2. Light requirements* of various plants

Light requirement	Subsistence	Growth
	fc	fc
Low	50	75-150
Medium	75-100	200+
High	200	500+
Very high	500	1000+

Source: Kaufman, P. B., T. L. Mellichamp, J. Glimn-Lacy, and J. Donald La-Croix. *Practical Botany.* 1985. Reston Publishing Company, Inc., Reston, Virginia. Used with permission of the authors.
* Light requirements are given for a 12-hour day. If receiving less than 12 hours of light, the fc will have to be increased. For example, plants normally requiring 200 fc for subsistence will need 400 fc if receiving only 6 hours of light. Most flowering plants require high to very high light sources.

intermediate light and fertilization programs (see Table 12.3). Plants are produced under partial shade varying from 40-80 percent. The desired shade is obtained by using plastic shade cloths of varying meshes. These are now commercially available in several shades. Fertilization is curtailed as the plant reaches marketable size and any remaining fertilizer can be leached prior to placement in the darker interior. Conductivity of a water extract (one part soil to two parts deionized or distilled water) of a sample collected just prior to placing the plant in an interior location ought to be in the 0.10-0.25 mmhos/cm range.

Supplemental Light

Periods of insufficient light for plant growth can occur anytime for interior plants in malls, homes, and other structures, but in greenhouses these are usually restricted to the winter months (in northern climates). Poor growing conditions due to low light intensity can be improved by the use of supplemental lighting. Several factors need to be considered before opting for supplemental light.

Although additional light may increase plant growth or quality, its use may not be justified economically. The relative costs of three sources of supplemental light used to produce several ornamental crops are given in Table 12.4.

The need for supplemental lighting or shading is dependent upon the plants photosynthetic activity. The photosynthetic activity of shade-loving C3 plants tapers off at 100 w/m^2 (3300 fc) and that of the sun-loving C3 plants at 200 w/m^2 (6600 fc). Therefore, there is no benefit in increasing light intensities of shade-loving plants, such as philodendron, beyond about 75 w/m^2 (2500 fc) and the sun-loving C3 plants beyond about 150 w/m^2 (5000 fc). The light intensity for plants with poor photosynthetic activity needs to be decreased to a range of about 75-150 w/m^2 (2500-5000 fc) by shading. On the other hand, increasing light intensity for the more active C3 plants to about 200 w/m^2 (6600 fc) can be helpful.

The light level can be varied depending on the result desired: 30 w/m^2 (1000 fc) for survival; 90 w/m^2 (3000 fc) for a 12-hour day for maintenance; 180 w/m^2 (3300 fc) for 6-8 hours for propagation; and 240 w/m^2 (8888 fc) for 8-16 hours for rapid growth in the greenhouse.

TABLE 12.3. Recommended light and nutritional levels for production of acclimatized foliage plants

Plant	Light	Shade*	Fertilizer**
	fc	%	Lbs N-P205-K20 a/yr
Aglaonema spp.	2000-2500	80	1200
Asparagus spp.	3500-4500	63	900
Calathea spp.	3000-3500	73	1200
Christmas cactus	3000-3500	73	1200
Croton, variegated	7000-8000	30	1800
Dieffenbachia spp.	3000-4500	63-73	1200
Dracaenas			
Corn plant	3000-4500	63-73	1200
Deremensis	3000-4500	63-73	1200
Marginata	5000-6000	55	1800
Others	3000-3500	73	1200
Ficus or fig			
Benjamina or weeping	3500-4500	63	1800
Elastica or rubber	5000-8000	30-55	1800
Fiddle leaf	5000-6000	55	1800
Nitida	3500-4500	63	1800
Norfolk island pine	5000-6000	55	1200
Palms			
Areca	5000-6000	55	1500
Bamboo, erumpens	5000-6000	55	1500
Bamboo, elegans	3000-3500	73	1200
Pilea spp.	2000-3500	73-80	600
Philodendron			
selloum	5000-6000	55	1800
spp.	3000-3500	73	1500
Pothos	3500-4500	63	1500
Sansevieria spp.	3500-4500	63	600
Scindapsus spp.	3500-4500	63	1500
Spathiphylllum			
Clevelandii	2000-3500	73-80	1200
White butterfly	3000-4500	63-73	1500
Zebra plant	1000-2000	80-90***	1500

*   Listed as percent actual shade.
**   Pounds of 18-6-12 Osmocote.
***   80 percent shade in the winter and 90 percent shade in the summer.

Source: From a table prepared by C. A. Conover and R. T. Poole, IFAS, Central Florida Research and Education Center, Apopka, FL 32703.

TABLE 12.4. Typical minimum costs of producing several ornamental crops under artificial lights

| Crop | Dry wt (g) | Minimum cost of production | | |
		Incandescent	Fluorescent	Metal arc
		(kilowatt-hours)		
Cut rose	7.0	4.0	2.0	0.7
Cut carnation	7.0	6.3	3.1	1.0
Cut gerbera	4.0	3.6	1.8	0.6
6" pot mum	40.0	35.7	17.8	6.4
4" petunia	5.0	4.5	2.2	0.7

Source: Kohl, H. and R. Evans. 1989. *Flower and Nursery Report.* Cooperative Extension, USDA, University of California, Berkeley, CA 94728.

Excess light can reduce net photosynthesis in many crops by increasing the temperature. The problem of heat-reduced net photosynthesis is especially serious in enclosed structures, where most of the solar radiation is converted to heat that has little chance to escape.

The benefits of a supplemental light program vary with the light source. The ordinary tungsten filament light bulb is a poor source of supplemental light if much light is needed, because it uses too much of the electrical energy to produce heat. Mercury tungsten lamps and mercury fluorescent reflector lamps do provide efficient light, with the mercury reflector lamps being more economical. Low-pressure sodium lamps (LPS) are efficient but they lack blue light. This deficiency can cause chlorosis of some chrysanthemums and abnormal growth of lettuce and tomato, but cool white fluorescent lamps (CWF) eliminate this type of chlorosis. Chlorosis with the LPS lamps increases as the ratio of ammonium to nitrate nitrogen (NH_4/NO_3) increases, indicating a relationship between utilization of NH_4-N and spectral quality. Normal plants can be grown with the LPS lamps, provided that about 40 percent of the total irradiance is supplied by white light. A disadvantage of the LPS lamps is the need for large reflectors. These reflectors often reduce low winter light to undesirable levels. High-pressure sodium lamps (HPS), although deficient in blue light, are superior to LPS lamps. Phytotoxicity symptoms, including epinasty (the downward bending of leaves), poor rooting of cuttings, abnormal leaf color, and leaf abscission of several plants have also been noted with these lamps. The abnormalities are also associated with N nutrition. The range of HPS lamps has been improved with the addition of rare-earth iodides to yield the MBI, HPI, HQI, or multivapor lamp (MVL), but because the output and life is less than the HPS lamps, they have been used less frequently. Finally, the economic benefits will be affected by the maximum efficiency and average life of the several sources that are given in Table 12.5.

MEASURING LIGHT INTENSITY

Light levels can vary tremendously in home or office interiors, shopping malls, greenhouses, shadehouses, or other protected structures depending upon location, season, time of day, weather, composition and cleanliness of the structure, and the extent of shade. Estimation of light intensity can be made by comparison with the fc of light emitted by different sources (Table 12.1), but such comparisons may be erroneous. On the other hand, measurement by light meters is very accurate.

Common light meters in the U.S. measure radiant energy in fc, but some instruments are calibrated in lux units. An incident light

TABLE 12.5. Comparative effectiveness and life of lamps used for artificial lighting

Lamp type	Maximum efficiencies lumens per watt	Average life hrs
Incandescent	24	1000
Halogen (Tungsten)	25	2000
High-pressure mercury	63	24000+
Fluorescent	100	20000
Metal halide	125	15000
High-pressure sodium	140	24000+
Low-pressure sodium	200	18000

Source: Courtesy Phillips Lighting Co., Somerset, NJ 08875.

meter capable of measuring up to 10,000 fc is desirable. The meter needs to be sensitive to 50 fc if it is necessary to maintain plants under low light levels as is the case in commercial, institutional, or residential buildings. In measuring the light level, the incident light meter is pointed with the sensing cell toward the light source. Descriptions of a few meters, as well as where they can be purchased, are listed in Table 12.6.

If a light meter is not available, an estimation of fc can be made using the built-in light meter in a single-lens reflex camera by using the formula:

$$\text{Foot candles} = \frac{20(f)}{SA}$$

where f = the f-stop number, S = the shutter speed in seconds and A = ASA film speed. A meter reading is taken 12 inches from an 8-inch by 11-inch sheet of white paper under lighting conditions to be measured and the camera is adjusted for proper exposure to obtain f and S.

A more accurate light measurement employs flat plate collectors as used by weather stations and a data logger to monitor the light periodically during the day, integrating multiple measurements from multiple sensors. A low-cost system consisting of the LI-COR Data Logger and two Li-190 PAR sensors that effectively measures light intensity over a prolonged period can be obtained from LI-COR, Inc., Lincoln, NE.

LIGHT DURATION

The duration of light of certain wavelength and irradiance affects seed germination, plant development, flowering responses, fruiting, and the development of storage organs.

The effect of day length, or rather the duration of darkness, is strikingly manifested in the onset of flowering of a number of plants. Three major groups of plants are categorized by their response to daylength:

TABLE 12.6. Light meters suitable for ornamental horticulture

Meter & model	Sensitive range fc	Area of use		Cost	Name & address of manufacturer or distributor
		Green & shade house	Building interior		
General Electric	0-10000	yes	yes	low	General Electric Cleveland, OH 44112
Gossen (Luna-Pro)	0-32000	yes	yes	med	Berkeley Mark. Co. Gossen Div. Woodside, NY 11377
Gossen (Panlux)	0-12000	yes	yes	high	"
Sekonic #L-398	0-12500	heavy shade	yes	med	Copal Corp. of America Woodside, NY 11377
Spectra (Candela)	0-30000	yes	yes	med	Photo Research Burbank, CA 91505

Source: Ingram, D. L. and R.W. Henley. *Nursery Laboratory Development and Operation.* IFAS, University of Florida. Circular 556.

1. *Short-day plants.* Plants, such as chrysanthemums and poinsettias, that initiate flower buds when the length of day (daylight) falls below a critical length.
2. *Long-day plants.* Plants, such as carnations and spinaches, that initiate flower buds when the day length is longer than the critical length.
3. *Day-neutral.* Plants, such as beans and tomatoes, that do not initiate flower buds in response to day length. Flower buds are started after the plants reach a certain size or have produced a certain number of nodes.

The flowering of a few plants takes place when a series of short days is followed by long days, while some others flower only if a series of long days is followed by short days. Complicating the whole process of flower induction, some plants need particular high or low temperatures to accompany the specific day lengths in order to initiate the process (Table 12.7).

Various light sources can be used for extending day length. Tungsten lamps, which are so inefficient for increasing light intensity, are satisfactory for extending day length because of the red light emitted.

TABLE 12.7. Photoperiod and temperature requirements for flowering of several plants

Temp. requirement	Photoperiod requirement		
	Day neutral	Short day	Long day
None	bean buckwheat corn cucumber Kentucky bluegrass	coffee pigweed sweet potato	carnation cvs clover foxtail oat cvs ryegrass cvs spinach cvs
High quantitative	fuchsia summer rice	cosmos chrysanthemum cvs	phlox
Low absolute	cabbage celery hydrangea	chrysanthemum cvs	carnation cvs ryegrass cvs spinach cvs sugar beet
quantitative	broad bean sweet pea vetch	chrysanthemum cvs	carnation cvs oat cvs winter barley winter wheat

Source: Adapted from Dennis, F. G. 1984. Flowering. In *Physiological Basis of Crop Growth and Development*. M. B. Tesar (ed.). American Society of Agronomy, Crop Science Society of America, Madison, WI 53711. Reproduced by permission.

Chapter 13

Temperature

Temperature is one of the more important factors affecting the rate of growth and is probably the most important factor regulating the natural distribution of plants. By altering a number of chemical and physical processes (diffusion, solution, evaporation, and reaction between chemical compounds), temperature plays an important role in internode elongation, leaf unfolding, and flower initiation. Temperature also affects plants indirectly by regulating the growth of microflora responsible for releasing nutrients from OM or fixing N from the air.

There are minimum and maximum temperatures at which plants can survive, although these are modified by the duration of a given temperature and the age and condition of the plant. There is a range of temperatures, usually 10-15°F above freezing, within which plants survive but grow slowly or not at all. Movement of temperature above the upper limit of this inactive range (base temperature) is accompanied by a rapid increase in the growth of most plants until the optimum range is reached, after which plant growth response will drop, usually very rapidly. (See Figure 13.1.)

The duration of temperature exposure can initiate or intensify the effects. The damage resulting from freezing temperatures or excessively high temperatures is greatly increased as the plant is exposed for longer periods. The exposure to cold periods during certain developmental stages can influence flowering. A number of temperate-climate plants need to be exposed to a given set of chilling temperatures before their dormancy can be broken and flowering can take place. There are also a number of species that have to be exposed to higher temperatures before they flower. In both cases, total hours of exposure to temperatures below or above a certain induction point are important in initiating flowering.

FIGURE 13.1. Typical temperature response curve, showing the rate of leaf formation as related to temperature

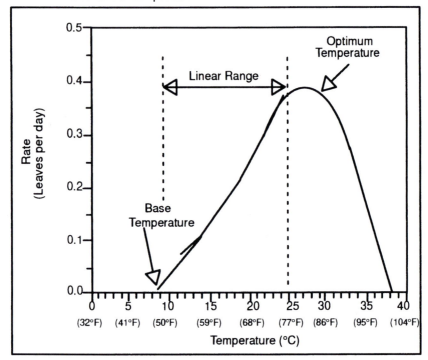

Source: Heins, R. D. 1990. Choosing the best temperature for growth and flowering. In *Greenhouse Grower,* 8:(4)57. Reprinted by permission.

The temperatures at which plants grow best usually occupy a relatively small range of responsive temperatures and this varies with different plants and stages of growth. For temperate-zone crops, this optimum range lies between about 55° and 90°F, with the base temperature at which plants respond favorably at about 45°F and maximum temperatures at which growth stops in the range of 100-120°F. Optimum temperatures for the germination of several crop seeds are given in Table 13.1.

Plants grow at the same rate as long as average daily temperatures within the linear range are the same. For example, plants will produce the same amount of growth if (1) they are grown at a

TABLE 13.1. Optimum temperature range for germination of vegetable seeds

Vegetable	Temp ° F	Vegetable	Temp °F
Asparagus	60-85	Okra	70-95
Bean	60-85	Onion	70-95
Bean, lima	65-85	Parsley	50-85
Beet	50-85	Parsnip	50-70
Cabbage	45-95	Pea	40-75
Carrot	45-85	Pepper	65-95
Cauliflower	45-85	Pumpkin	70-90
Celery	60-70	Radish	45-90
Chard, Swiss	50-85	Spinach	45-75
Corn	60-95	Squash	70-95
Cucumber	60-95	Tomato	60-85
Eggplant	75-90	Turnip	60-105
Lettuce	40-80	Watermelon	70-95
Muskmelon	75-95		

Source: Lorenz, O. A. and D. N. Maynard. 1988. *Knott's Handbook for Vegetable Growers*. Third Ed. New York: John Wiley & Sons, Inc. Reprinted by permission of John Wiley & Sons, Inc.

TABLE 13.2. Response of plants to the difference between day and night temperatures

Large responses		Small or no response
Bean, snap	Hypostes	Aster
Celosia	Impatiens	Hyacinth
Chrysanthemum	Petunia	Marigold, French
Corn, sweet	Poinsettia	Narcissus
Dianthus	Portulaca	Platycodon
Fuchsia	Rose	Squash
Gerbera	Salvia	Tulip
Lily	Snapdragon	
Asiatic	Tomato	
Easter	Watermelon	
Oriental		

Source: Heins, R. and J. Erwin. 1990. Understanding and applying DIF. *Greenhouse Grower*, 8:(2)75. Reprinted by permission.

12-hour day temperature (DT) of 80°F and a 12-hour night temperature (NT) of 60°F; (2) at a DT of 65°F and an NT of 75°F; or (3) at a constant temperature of 70°F. The maximum rate of growth is obtained if DT is the same as NT, but an equivalent average of DT and NT will give the same rate as obtained from uniform DT and NT. Differences in DT and NT, however, are responsible for variations in the height of sensitive plants (Table 13.2), with plant height increasing as DT increases, but decreasing as NT increases.

The market life of harvested commodities is dependent on maintaining of ideal temperatures, which vary with different commodities (see Table 13.3). Those requiring cool temperatures need to be cooled as rapidly as possible.

HEAT-INDUCED STRESS

Exceeding optimum temperatures for growth induces heat stress. Growth slows and if heat stress is continued for long periods can come to a complete halt. Yield, plant appearance, and fruit set will be negatively affected.

Heat stress can be reduced or eliminated in the field by proper site selection, the use of mulches, the addition of sufficient soil moisture, and the frequent application of overhead irrigation. In addition to these measures, proper shading, ventilation, and cooling with evaporative fans offer plants relief from heat stress in closed structures.

COLD-INDUCED STRESS

Cold-induced stress takes two forms: (1) a chilling injury that occurs above freezing; and (2) a freezing injury that takes place at or below freezing.

Chilling Injury

Chilling injury affects tropical and herbaceous plants not adapted to withstand freezing temperatures. Harvested commodities as well as growing plants can be affected. It can occur as air temperatures

TABLE 13.3. Optimum storage temperatures* for harvested commodities

Temp °F	Fruits	Vegetables		Flowers	
32*	Apple**	Artichoke	Garlic	Carnation	
	Apricot	Asparagus	Leafy	Chrysanthemum	
	Berries***	Bean	greens	Iris	
	Cherry	Brussels	Horseradish	Lily-of-the-valley	
	Date	sprouts	Kale	Dry rose	
	Fig	Cabbage	Kohlrabi	Sweetpea	
	Grape	Carrot	Leek	Tulip	
	Kiwi	Cauliflower	Lettuce		
	Nectarine	Celeriac	Mushroom		
	Peach	Chard	Onion		
	Pear	Chicory	Parsley		
	Persimmon	Collard	Parsnip		
	Plum	Corn	Rutabaga		
	Prune	Endive	Salsify		
	Quince	Escarole	Spinach		
			Turnip		
			Watercress		
32-35	Apple cvs	Asparagus		Allium	Hyacinth
	Orange	Cantaloupe****		Aster	Narcissus
		Water		Bouvardia	Cymbidium
		chestnut		Crocus	orchid
				Freesia	Ranunculus
				Gardenia	Rose in
				Gerbera	preservative
35-45	Apple cvs	Bean		Acacia	Marigold
	Avocado*****	green		Alstroemeria	Poppy
	Cranberry	lima		Anemone	Phlox
	Guava	Cantaloupe******		Aster	Primrose
	Orange	Pea,		Bird-of-	Protea
	Pomegranate	southern		paradise	Snapdragon
	Tangerine	Squash,		Buddleia	Statice
	Mandarin	summer		Calendula	Stock
		Tamarillo		Calla	Strawflower
				Candytuft	Sweet
				Columbine	william
				Cornflower	Violet
				Dahlia	Zinnia
				Daisy	Florists
				Delphinium	greens
				Gladiolus	
				Gypsophilia	
				Heather	
				Lily	
				Lupine	

TABLE 13.3. (continued)

Temp °F	Fruits	Vegetables		Flowers
45-55	Avocado	Cucumber	Pepper	Heliconia
	Carambola	Eggplant	sweet	Cattleya orchid
	Lemon	Melons	Pumpkin	Sweet william
	Lime	casaba	Squash	Bird-of-paradise
	Papaya	crenshaw	winter	
	Passion	honeydew	summer	
	fruit	water	Taro	
	Pineapple	Okra	Tomato	
			ripe	
55 or	Banana	Ginger	Potato	Anthurium
more	Grapefruit	Jicama	sweet	Ginger
	Mango	Watermelon	Tomato	Vanda orchid
	Plantain		green	Poinsettia

* Or below (but above freezing point).
** Apple cultivars vary as to ideal storage temperatures.
*** Except cranberries.
**** Fullslip.
***** Ripe.
****** 3/4 Slip.
Source: Kasmire, R. J. and J.F. Thompson. 1992. Selecting a cooling method. In *Postharvest Technology of Horticultural Crops*. A. A. Kader (ed.). Division of Agriculture and Natural Resources, University of California, Oakland, CA 94608-1239.

fall below 55°F or if cold water is applied to the leaves of growing plants. Mild damage can result at temperatures of 41-55°F, with serious damage occurring in the range of 33-41°F. Plants of the different crop groups that are readily injured are: *agronomic* (cotton, corn, rice, millet, and sorghum; *fruit* (avocado, banana, citrus, mango, and pineapple); *vegetable* (okra and tomato); and *ornamental* or *foliage* (Aglaonema, Fittonia, hibiscus, philodendron, Sansevieria, Scindapsus, and Syngonium).

Damage resulting from chilling can manifest itself to varying degrees, from practically no symptoms except for a slowed growth, to increasingly serious symptoms of: (1) leaf yellowing and loss; (2) various lesions (water soaked, pitted, and sunken areas); (3) browning, pitting, and abnormal ripening of fruits; (4) wilting and browning of part or the entire plant; and (5) accelerated death of the plant.

Freezing Injury

This type of injury occurs as temperatures reach and fall below 32°F. Frost, or the formation of ice crystals at leaf surfaces, occurs as the dew point (temperature at which dew forms) is reached at freezing temperatures.

Symptoms of freezing damage include desiccation or "burning" of the leaves, loss of leaves, water-soaked areas, necrotic spots, split bark of stems and branches, weakened root system, and death of plant parts of the entire plant. Wilting or desiccation often occurs because of impaired root systems and damage to conducting tissue. Some of these symptoms will not appear until days or even weeks after the freezing damage, making it desirable to wait for some time after the freeze to assess the damage.

The temperatures at which damage takes place varies according to species, the type of tissue, condition of the plant, the presence of ice-nucleating bacteria or ice-nucleating agents intrinsic to plant tissue, and the duration of temperatures below freezing. Usually tender young tissue is more prone to damage than older developed tissue, although there are young stages in some plants, such as tomato, which allow for cooling below 32°F for short periods without freezing. Flower buds are more sensitive to cold than vegetative buds or stem tissue. Open flowers are usually very sensitive. Root tissue is more sensitive than stem tissue, but is usually protected by the insulation of the earth. The freezing points of several commodities are given in Table 13.4.

Reducing Cold-Induced Stress

Cold-induced stress can be reduced by proper plant selection, proper site selection, increasing plant hardiness, proper use of water and fertilizer, and by manipulating temperatures.

Temperature Manipulation for Reducing Cold-Induced Stress

Utilizing temperature control to reduce cold-induced stress is made easier by understanding how heat is transferred. Heat is transferred from one object to another by conduction, convection, and

TABLE 13.4. Freezing points of several fruits and vegetables

Commodity	Range in Freezing Points (°F)
Avocado	29.9-31.5
Bean, snap	29.8-30.7
Cabbage	29.8-30.4
Eggplant	30.2-30.6
Lettuce	30.8-31.3
Lime, Key	28.0-28.5
Lime, Persian	28.0-29.5
Mango	29.0-30.5
Okra	28.3-28.7
Papaya	30.0-30.5
Tangelo, 'Orlando'	27.0-30.1
Tangerine, 'Dancy'	28.3-30.1
Squash, 'Yellow Crookneck'	30.1-30.8
Tomato, 'Homestead'	29.9-30.8

Source: Whiteman, T. M. 1957. Freezing Points of Fruits, Vegetables, and Florists Stocks. U.S. Marketing Research Report 196.

radiation. *Conduction* moves heat from one molecule to an adjacent one, increasing as molecules are closer and the material denser. *Convection* moves heat by mass motion of molecules in air and water. Convection is responsible for cold damage in low areas as the lighter, warmer air moves upward to be replaced by the colder, heavier air. *Radiation* moves heat from a warmer to a colder object without benefit of a transfer medium. All living and inanimate objects radiate heat, moving from the warmer to the colder object in long wavelengths that have little penetrating power. These waves travel in straight lines until they strike an object which can absorb, deflect, or reflect them. The earth is warmed during the day by radiant heat transfer from sunlight, which is present as short, highly penetrating wavelengths. The heat can be useful at night as it is radiated back to plants above the soil or to warm the air immediately above it. Unfortunately, a cold night with little or no cloud cover allows the heat to escape to the upper atmosphere where it does little good for the plants.

Cultural Practices and Structures

In climates having markedly fluctuating temperatures, but in which the temperatures do not drop low enough to freeze wet soil, much can be done to lessen the risk of cold-induced stress by making the most of heat generated from the sun during the day. This requires (1) removing mulches or avoiding cultivating moist soil so that heat can be absorbed during the day and can be released during the cold night; (2) applying insulating materials late in the day to conserve heat; and (3) using row covers to trap heat.

Conservation of the sun's heat so that it can be released during the cold night is greatly aided by the use of structures covered by glass or plastic. Air inside a house covered with a single sheet of plastic is about 5°F warmer than the air outside on a cold night. While this may be enough to lengthen the growing season or to help grow cold-tolerant crops during frosts or mild freezes, it usually is insufficient for most crops to be grown successfully off-season. Water from deep wells circulated through PVC pipes imbedded in sand or concrete on which plants are placed has been sufficient in a number of cases to keep plants above critical temperatures. Passive systems that trap the sun's heat during the day and store it in various media to be circulated through the house during the night increase the number of crops that can be successfully grown during the winter period.

Covering these structures during the night with various insulating materials reduces heat loss and increases the effectiveness of the structure. The use of double layers of plastic film that are air separated or extra single or double layers of plastic (air separated) placed over glass or fiberglass houses can materially reduce heat loss. Blankets at sidewalls and greenhouse ends, and the use of polyurethane isocyanurate or polystyrene boards to 18 inches below the foundation have also been helpful. Lightweight materials with good insulation that can be quickly drawn above the plants are effectively used to reduce heat losses during the night. (Protected structures may need to be cooled at times because of rapid heat buildup during sunny weather.)

SUPPLEMENTAL HEAT

Supplying supplemental heat to enclosed structures greatly increases the number of plants that can be grown off-season. Electric cables embedded in the soil are useful in the germination and early growth of several plants grown in structures that are not normally heated. Commercial production of several crops in certain regions is made possible by using waste heat from power plants. Heaters that provide radiant and/or convectional heat using wood, gas, coal, or oil as sources of energy are used to produce a wide range of crops.

Using fans in conjunction with heating devices helps to insure proper heat distribution and uniform heat through the house. Adequate numbers of low-pressure, high-volume fans need to be properly placed. Dual-control thermostats for every fan help in automatic control. Drawing outside air through a tubular plastic distribution duct helps provide adequate ventilation without injuring plants. When temperatures are above 50°F, air must be drawn through side vents or some means other than tubular ducts.

Supplemental heat from oil or gas burned in smudge pots or orchard heaters is also used outdoors to reduce cold damage. Placed under the tree's canopy or where much of the heat can be held close to the tree can give effective cold protection, although such practices are being seriously curtailed in many places because of environmental concerns.

USING WATER FOR PROTECTION AGAINST COLD DAMAGE

Adding sufficient water just prior to the cold spell helps a plant withstand the dehydration effects induced by cold. However, the application of water must not encourage new growth, which is far more vulnerable to cold damage. Actually, mild drought stress induced about a month before cold weather sets in helps the plant resist cold damage by slowing growth and inducing dormancy.

The application of water during temporary cold spells in a fluctuating climate can result in considerable cold protection. The water can be applied under or alongside the plant, directly to the plant with overhead irrigation, or as a fine mist to form a fog. The protection obtained is dependent on such factors as the temperature of the water, the amount of water applied, the wind velocity, and radiation effects.

Application of water by flooding, sprinkling, and fogging can be helpful in reducing damage from cold temperatures. The higher-than-air temperature of water from deep wells can in itself help maintain a higher leaf temperature. This elevated temperature is aided by the 232 British Thermal Units (BTU) released per gallon of water as it is cooled from 60°F to 32°F. (A BTU is the amount of heat required to increase the temperature of a pound of water by 1°F.) Water applied in furrows or under trees for the entire cold period (until temperatures rise several degrees above the freezing point), can elevate leaf temperatures enough to make a substantial difference in damage control, especially for temperatures near or just below the freezing point.

The application of water directly to plants by overhead sprinklers can reduce cold losses during much colder temperatures. Sprinkling has been useful for cold protection of certain fruits, vegetables, ornamentals, and fruit trees. Sufficient water must be added to keep the leaves wet as the evaporation of the water takes heat from the leaves and can increase injury. Sprinkling must start before critical temperatures are reached. Once the leaves are wetted, water application must continue until the danger of frost or freezing passes, as indicated by thawing or as the wet bulb temperature rises above the freezing point. The continuous application of the warmer water helps maintain adequate leaf temperature and once ice forms, large quantities of heat are released (1200 BTU per gallon of water). The latent heat released by freezing prevents leaf temperatures from falling below 32°F, as a mixture of ice and water has a temperature slightly above 32°F. This is sufficient to prevent the freezing of many fruits and vegetables (see Table 13.4). To be effective, the entire upper part of the plant must be covered with the ice mixture. The amount of water needed increases with low dew points and as the wind increases. The amounts of water needed with different temperatures and wind speeds up to 8 miles per hour are given in Table 13.5.

Applying water at a rate of 0.5-1.5 gal/min by means of micro-sprinklers placed 2-3 feet above the ground in the scaffold branch

TABLE 13.5. Recommended application rates of water for cold protection with different wind and temperature conditions

Lowest temperature expected	Wind speed in miles per hour		
	0-1	2-4	5-8
°F	Inches of Water to be Applied per Hour		
27	0.10	0.10	0.10
26	0.10	0.10	0.14
24	0.10	0.16	0.30
22	0.12	0.24	0.50
20	0.16	0.30	0.60
18	0.20	0.40	0.70
15	0.26	0.50	0.90

Source: Harrison, D. S., J. E. Gerber, and R. E. Choate. 1974. Florida Technical Circular 348.

area can produce high survival rates for citrus trees exposed to hard freezes. Although canopy survival may be poor, a sufficient percentage of scaffold branches usually survive to permit about 95 percent tree regeneration, whereas almost 100 percent of the trees that receive no irrigation may be killed.

There is a possibility of inducing cold stress by applying cold water to warm plants. This situation arises most often in greenhouses, where water of 55°F or less is applied to tropical foliage plants, a number of which may show yellow leaves that readily defoliate a week or two after they have been watered. This problem is readily avoided by using warm water or water that is preheated to about 60°F.

The Use of Fog

Application of water to produce a fog benefits both from the heat released from the cooling water and the lowered loss of heat due to radiation as the fog acts as an effective cloud cover. Fogging can be highly useful in enclosed structures, where it can raise air temperatures as much as 9°F. To have maximum effectiveness, fog must be uniformly distributed. It is best created by the use of high pressure, low volume systems. Fog produced by flooding is useful outdoors where temperatures can be raised 4°F if the water temperature is at least several degrees warmer than the ground surface. Application of water to reduce chilling effects is most useful on still nights, when there is less radiation loss.

MEASURING TEMPERATURE

Utilizing temperature to maximize crop production requires accurate measurement by thermometers. Thermometers suitable for the purpose may vary from the simple hand-held type or the more sophisticated instrument capable of recording maximum-minimum temperatures, to those capable of sounding an alarm if temperatures fall too low or rise too high, or even to the sensaphone types that will automatically respond to programmed alarm limits by phoning temperature information to preset numbers.

Recorders, some with humidity-measuring ability, are useful for maintaining records. Thermostats controlling heating or cooling devices can make the automatic control appreciably simpler. Automatic thermostat control propagation mats for germinating seed are very helpful. Greenhouse controllers that regulate heating and cooling devices automatically for an entire greenhouse for both night and day conditions are available. These controls can be programmed into a computer to effectively regulate temperature by utilizing solar energy, heating equipment, cooling devices, and humidity control.

In order to obtain accurate information from thermometers, the thermometer must be calibrated and placed in a position that accurately reflects effective temperature. Thermometers can be calibrated by immersion in an ice bath, the temperature of which is 0°C or 32°F. At least some of the thermometers need to be placed close to growing plants so that they accurately reflect growing conditions. Thermometers should not be placed so that they can gain or lose radiant heat to or from such sources as motors, cold or hot walls, or people. Errors due to radiant heat loss or gain can be avoided by the use of a radiant heat shield or by having neutral objects between thermometers and radiant heat sources.

Chapter 14

Atmosphere

The atmosphere consists of the normal gases (carbon dioxide, nitrogen, oxygen, water vapor, and ethylene) and several gases which are largely the by-products of industrialization. The concentrations of carbon dioxide, oxygen, and nitrogen remain fairly constant at given temperatures and elevations, although carbon dioxide appears to be on the rise due to increased combustion of OM. Water vapor content varies considerably with location and air temperature.

Also on the rise is a group of gases, products of industrialization, that are causing considerable concern because of their harmful effects on a great variety of plants. Damage is usually exhibited in or near urban areas, although the use of tall smokestacks has enabled some of these gases to travel great distances. Damage to sensitive growing plants or harvested commodities caused by combustion gases can be severe in greenhouses, packing sheds, storage areas, and/or trucks unless these spaces are properly ventilated.

Many harvested commodities have a longer shelf life if they are transported or stored in an atmosphere of reduced oxygen and elevated carbon dioxide. The modified atmosphere (MA) can be obtained passively as oxygen is reduced and carbon dioxide increased by respiration. Use of certain packaging films can speed the process. More often, better results are obtained by use of controlled atmosphere (CA), which changes the content of oxygen and carbon dioxide.

Better growth of crops and longer life of harvested products are possible by controlling certain aspects of the atmosphere. Control to some extent is made possible by the use of instruments capable of diagnosing the various atmospheric constituents, but it is important to know how the various components of the atmosphere affect plants. A description of the various constituents and the manner in which they affected production, as well as means of measurement, are outlined below.

CARBON DIOXIDE

Carbon dioxide (CO_2), which comprises 0.033 percent of the atmosphere, is the source of carbon (C) in plants. It is the most abundant element in plants, being an integral component of all OM and comprising about 45 percent of its dry weight.

CO_2 Enrichment

Under normal sunlight conditions, lack of sufficient CO_2 in greenhouses can limit photosynthesis more frequently than inadequate light. The CO_2 content of tightly closed greenhouses may drop below normal atmospheric concentration. Enrichment with CO_2 under such conditions has proven profitable for growers in Holland and England. A recent study* by M. M. Peet and D. H. Willits of North Carolina State University has shown that such enrichment can be employed profitably with cucumbers, and possibly with lettuce and tomatoes as well. Cucumber and lettuce yields are enhanced even by short periods of enrichment, provided CO_2 concentration is high enough. Peet and Willits determined that the application of 14,400 ppm of CO_2 for one hour gave about the same yield of cucumbers as 1000 ppm for 14.4 hours or 3000 ppm for 4.8 hours. Tomato yields do not increase as markedly with short periods of enhancement, but they do show substantial increases for all day enrichment.

Carbon dioxide enrichment has been used in a limited way for some other high-priced crops, such as flowers and foliage plants, produced in protected structures. For maximum effects, the temperature, fertilizer, water, and even light may need to be increased along with the CO_2.

Increasing the CO_2 surrounding the plant from the normal concentration of about 300 ppm (0.03 percent) up to about 900 ppm can

---

*Willits, D. H. and M. M. Peet. 1989. Predicting yield responses to different greenhouse CO_2 enrichment schemes: Cucumbers and tomatoes. *Agriculture and Forest Meteorology*, 44:275-293.

increase the photosynthetic activity of many crops. At the higher level of CO_2, the light intensity has to be about three times that which is suitable at 300 ppm CO_2. Whereas 500 fc may be satisfactory for some crops grown at ambient CO_2 concentrations, about 1500 fc are needed to obtain a desirable response to the extra CO_2. This is no problem for plants exposed to full sunlight, which supplies about 10,000 fc, but light may need to be supplemented under certain conditions.

Increasing the CO_2 of irrigation water used with black poly mulch has been shown to increase the CO_2 content of the soil and to produce increased tomato yields. The pH of the water drops about 2 units and that of the soil about 1 unit, which could be a plus if high pH water is to be used on soils already high in pH. Such decreases in pH, however, could cause problems with lower pH soils that contain potentially high levels of Al or Mn.

The use of CO_2 enrichment in protected houses may be possible with alternative closed-loop systems, but such systems do not offer enough cooling for warm, humid areas. It may be possible to use heat-pump-type air conditioners for closed-loop greenhouse cooling and this approach is now being investigated.

For improved shelf life of harvested commodities, the 0.033 percent concentration of CO_2 in the air has to be increased by a couple of percentage points for a number of vegetables, and by 10-15 percent for cantaloupes, green onions, and spinach. The high level of CO_2 needed in transport vehicles or storage areas can be obtained by the use of dry ice or, as is more common, from pressurized gas cylinders. In small areas, such as covered pallets or packages, the needed CO_2 level can be reached gradually by respiration. The process can be hastened by proper selection of film or coverings.

Measurement of CO_2

The concentration of CO_2 in protected structures must be monitored to maintain desirable levels for growth and to determine whether CO_2-generating equipment is working satisfactorily. The gas can be measured (300-5000 ppm) manually by using gas detector tubes or gas analyzers that incorporate microprocessors. Some gas analyzers incorporate time clocks and are capable of displaying CO_2 concentration in ppm or as a curve of ppm/time. The gas analyzer can stand alone, or it can be set with a control relay to activate a CO_2 generator or combined with a computer for more complete evaluation of growth parameters.

Laboratory methods using an infrared CO_2 analyzer or gas chromatography are more accurate than on-spot methods, but these have the disadvantages of special sampling procedures and the extra time needed to obtain results.

OXYGEN

Oxygen (O_2) makes up about 21 percent of the atmosphere's volume and 45 percent of a plant's dry weight. It is essential for respiration, by which energy is made available for various plant processes. The concentration aboveground is almost never limiting for plant growth, but O_2 in the soil can be so reduced by flooding, compaction, and poor soil structure that respiration taking place in the roots can be seriously affected. The importance of soil O_2 and diagnostic approaches that can be used to avoid shortages in soil were considered in Chapter 2.

Oxygen reduction in storage rooms and transport vehicles is an integral part of commodity shelf-life enhancement. The normal content of 21 percent O_2 in the air must be reduced to less than 5 percent in order to prolong the shelf life of a number of commodities. This may be done passively in closed containers (packages, sealed pallets) or areas, as O_2 is naturally consumed in the respiration process. In recent years, there has been an increasing use of CA methods, in which O_2 is lowered by purging with nitrogen (N_2).

Measurement of Oxygen

On-spot measurement of O_2 is possible with methods similar to those for CO_2, i.e., gas analyzers, some of which are portable, and detector tubes. Laboratory analysis is made possible by means of polarographic, electrochemical, and paramagnetic O_2 analyzers, and by gas chromatography.

WATER VAPOR

The water vapor content, or humidity, of the atmosphere is very small (0.4-1.5 percent of the weight of air), but varies considerably, decreasing as elevation is increased. The content of the lower atmosphere also varies with location, with highest concentrations over equatorial and monsoon regions and the lowest concentrations over subtropical deserts.

Relative humidity (RH) is the ratio of the amount of water (weight) held by air at any given temperature and pressure to the amount that can be held. At 100 percent RH, the air is saturated with water. Relative humidity is only slightly affected by atmospheric pressure, but it is greatly influenced by temperature since a great deal more water is held by warm as compared to cold air. Generally, the amount of water that can be held by air is doubled for every 20°F rise in temperature . There is about a 2 percent decrease in RH for each 1°F drop in temperature. Condensation of air moisture develops as warm humid air temperatures are dropped to the point at which moisture exceeds the lower holding capacity of the colder air. At temperatures above 32°F, dew is formed (dewpoint); at temperatures below 32°F, frost is formed (frost point). While high RH may be beneficial for crop growth and may prolong the shelf life of many commodities, condensation of water vapor either as dew or frost is generally harmful.

Relative humidity affects plants in several ways. Besides slowing heat loss, high RH reduces water loss from soils and plants (transpiration), making limited water supplies more efficient. Plant growth is favored unless the high RH is the result of prolonged rainy periods. Such periods are often associated with less solar radiation and lower soil nutrient levels as a result of leaching. High RH, by slowing the evaporation of foliar sprays, increases the uptake of nutrients from the spray. High levels of humidity aid in rooting cuttings and reducing damage from F. But high RH, by lowering transpiration, can negatively affect the uptake of N, P, K, Ca, and Mg, and possibly the micronutrients as well. The reduction is often of minor importance, but movement of Ca, an element that moves primarily in the transpirational stream, can be lowered enough to cause Ca deficiencies in certain plants ("tipburn" of lettuce, escarole, and English pea). Although high RH favors the increase of several bacterial and fungal diseases, it reduces damage from several mites. Humidity control often is the key for controlling a number of diseases in protected houses.

Regulating RH

High RH can be reduced in enclosed structures by increasing temperature and ventilation. The simplest way of reducing RH is by ventilation with air that has a lower vapor content. Heating by itself is not an efficient way of reducing RH in greenhouses, but in combination with the opening of vents in the spring and fall months it is very effective. The increased temperature allows more moisture to be held (lower RH), and increases the rate of air movement, allowing a greater exchange with outside air.

Although condensation of water on leaf surfaces during the winter is avoided with proper heating, it may form on glass or plastic coverings. Usually, it is not a problem in glasshouses, but it can be troublesome in plastic houses, where fewer roof openings mean less exchange of air. The problem is reduced by the use of polyethylene films with modified surface tension or by using two layers of plastic.

RH in greenhouses can be increased by wetting down walks, by using evaporative cooling, or by using fog. Increases in humidity from wetting walks is of relatively short duration. The other methods produce longer-lasting results and also provide advantages in controlling temperature.

Humidity for harvested commodities can be increased by wetting down storage areas, by packaging, and by using films to cover pallets. Low RH, which is desirable for curing several commodities, may occur naturally in certain regions, but often it is obtained by forcing air heated by the sun or by combustion of fossil fuels over the commodity.

Measuring RH

Relative humidity in a range of 20-100 percent can be accurately measured by the combined use of dry-bulb and wet-bulb thermometers. (The RH is higher as the wet bulb temperature is closer to that of the dry bulb). The thermometers must be accurate and have a sensitivity of at least 0.5°F. The RH is obtained from a psychometric chart (Figure 14.1) by finding the wet-bulb and dry-bulb temperatures on their respective lines in the chart and noting the RH where these lines meet. Pocket calculators are available that automatically provide the RH from imputs of wet- and dry-bulb readings.

FIGURE 14.1. Relative humidity from wet- and dry-bulb thermometers (F)

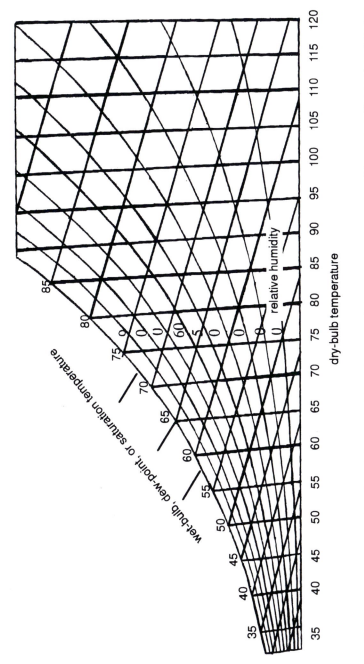

Note: To obtain relative humidity (RH), follow straight line for dry-bulb temperature upward until it crosses diagonal straight line representing wet-bulb temperature. Relative humidity is read from curved line at junction point.

To obtain meaningful readings with the wet-bulb thermometer, it must also be shielded from radiant sources, have a cotton wick wetted with distilled or deionized water surrounding the mercury reservoir, and subject the wick to air movement at sufficient speed. This rate of air movement is readily obtained by rapidly rotating the hand-held sling psychrometers. Wall-mounted hygrometers using both wet- and dry-bulb thermometers, require some external source of air movement, usually in the form of battery-operated fans which readily exceed the 500 feet per minute-rate.

Air speed past the wick must exceed 500 feet per minute. Use of sling or motor-blower psychrometers allows for sufficient air speed. Wall-mounted hygrometers, using both wet- and dry-bulb thermometers, require some external source of air movement for accurate readings.

Humidigraphs giving direct readings of RH are also available. These depend on expansion and contraction of a hygroscopic sensor. Devices that record both temperature and RH are available and are useful for monitoring RH and temperature in storage areas and transport vehicles. A modular system based on humidity and temperature sensors that automatically adds fog as needed to maintain a preset RH is available from Mee Industries Inc., El Monte, CA 91731.

AUTOMATIC CONTROLS

A number of devices are available that monitor environmental conditions and can be programmed to activate various control mechanisms to provide ideal conditions for greenhouses and other enclosed structures. Some of the conditions that can be controlled are: light, temperature, humidity, and CO_2 concentration. A few systems now in use are manufactured or distributed by: Com Corp., Irvine, CA 92714; Lander Control System, Inc., Guelph, Ontario, Canada NIH 6H8; Micro Grow Greenhouse Systems, Inc., Temcula, CA 92591; Wadsworth Control Systems, Arvada, CO 80002, and En Tech Control Systems, Inc., Montrose, MN 55363.

Chapter 15

Pollutants

Several air pollutants affect production of crops, their quality, and the length of their shelf life. Pollutants can also cause serious economic losses by adversely affecting appearance, even though there may be relatively little reduction in yield.

Generally, there are two types of pollution: (1) that which affects all plants, and results largely from man-made gases, and (2) that which occurs in enclosed structures, where plants or commodities can be exposed to both general atmospheric pollution and that produced in the structure. The latter type of pollution can be extremely damaging because the enclosures facilitate the buildup of concentrations to lethal levels.

ATMOSPHERIC POLLUTANTS

Nitrogen

Most of the nitrogen (N) in the atmosphere takes the form of gaseous N_2 and is naturally present. It makes up about 78 percent of the total composition of the atmosphere. But the atmosphere may also contain small quantities of ammonia (NH_3), produced in urban areas, and nitrates (NO_3), derived from electrical discharges during storms. These materials are brought down to earth during rains and can supply some of the N for plants.

Damage from N Pollutants

Damage from free NH_3 is primarily confined to enclosed structures and is covered later in the chapter.

In urban areas, the atmosphere can contain nitrogen dioxide (N_2O_2) produced from hot combustion sources such as open fires, furnaces, and automobiles. Relatively low concentrations of about 2-3 ppm can be harmful to sensitive crops such as beans and tomatoes. Symptoms of damage appear as bleaching of marginal and interveinal leaf tissue and are similar to sulfur dioxide (SO_2) injury.

Peroxyacetl nitrate (PAN), an N compound which results from combustion of organic fuels, is present in the atmosphere in relatively large quantities near urban areas and is responsible for causing considerable damage to a wide variety of plants. The damage attributable to smog appears to be partly due to this compound. Even in concentrations of as little as 0.2 ppm, PAN has been known to damage sensitive plants. (See Table 15.1.) Typical symptoms are stunting and early maturity, which may be accompanied by bronzing, glazing, and silvering of broad-leaf species and the bleaching of grass leaves. The grass leaves may also have tan to yellow bands. Damage to conifers shows up as yellowing of the needles.

Chlorine and Chlorides

Small quantities of chlorides (Cl) are borne from salt water by wind, but these are probably of little concern except in areas very close to the sea. A more serious problem appears to result from smokestack emissions of gaseous chlorine Cl_2, which combines with water to produce hydrochloric acid (HCl).

Chlorine Damage

Exposure for one hour to concentrations of 0.46-4.67 ppm Cl_2 have produced symptoms on sensitive plants. (See Table 15.1.) The symptoms appear as necrotic areas between veins near the leaf margins of broad-leaf plants and progressive streaking toward the main vein in the upper region of grass leaves.

Fluorides

Hydrogen fluoride (HF), emitted from the smokestacks of aluminum, ceramic, and fertilizer (particularly phosphate) plants, can

TABLE 15.1. Plants affected by atmospheric pollutants

A. Sensitive Plants

PAN	HCl	HF	O₃	C₂H₄	SO₂
African	Alfalfa	Alfalfa	Alder	Avocado	Alfalfa
violet	Apple	Apple	Alfalfa	Banana	Apple
Alfalfa	Ash	Apricot	Apricot	Calceo-	Apricot
Aster	Basswood	(Chinese)	Ash	laria	Ash
Beets	Cherry	Azalea	Aspen	Canta-	Aspen
(sugar,	Corn	Barley	Aster	loupe	Aster
table)	Dogwood	Begonia	Avocado	Honeydew	Bachelor's
Bluegrass	Maple	Blueberry	Barley	Larkspur	button
(annual)	Mustard	Box elder	Beans	Lemon	Barley
Carnation	Oak	Buckwheat	(green,	Mango	Beans
Celery	Onion	Canna	pinto)	Orchid	(broad,
Chrysan-	Pine	Cherry	Beets	(dendro-	garden)
themum	(white)	Chives	(sugar,	bium)	Beech
Dahlia	Radish	Citrus	table)	Orange	Beets
Dill	Sunflower	Corn	Begonia	Rose	(sugar,
Endive	Sweetgum	Crocus	Bentgrass	Snapdragon	table)
Escarole	Tobacco	Cyclamen	Birch	Tomato	Begonia
Fennel	Zinnia	fir	Bluegrass		Birch
Fir		(Douglas)	(annual)		Black-
(Douglas)		Freesia	Box elder		berry
Lettuce		Gladiolus	Bridal		Bluegrass
(romaine)		Grape	wreath		(annual)
Lilac		Iris	Broccoli		Broccoli
Muskmelon		Jerusalem	Brome-		Brome-
Mustard		cherry	grass		grass
Oat		Larch	Brussels		Brussels
Orchid		(western)	sprouts		sprouts
Pepper		Lily	Carnation		Buckwheat
Petunia		Mahonia	Carrot		Cabbage
Pines		Maple	Catalpa		Carrot
(Coulter,		Mulberry	Celery		Catalpa
Monterey		Oak	Chicory		China
ponderosa)		Palm	Chinese		aster
Poinsettia		(Queen)	cabbage		Columbine
Potato		Peach	Chrysan-		Cosmos
Salvia		Peony	themum		Cotton
Small grains		Pines	Citrus		Crabapple
Spinach		(white,	Corn		Cucumber
Sunflower		yellow)	Crabapple		Dahlia
Swiss chard		Prune	Dahlia		Dandelion
Tobacco		(Italian)	Dill		Eggplant
Tomato		Poinsettia	Eggplant		Elm
		Poplar	Endive		Endive
		Spider	Fuchsia		Fir
		plant	Grape		(Douglas)
		Sorghum	Grasses		Forsythia
		Sweet	Hemlock		Hawthorn
		potato	Honey		Larch
			locust		Lettuce

TABLE 15.1. (continued)

A. Sensitive Plants

PAN	HCl	HF	O$_3$	C$_2$H$_4$	SO$_2$
		Sycamore	Larch		Mulberry
		Transvaal	Lilac		Mustard
		daisy	Linden		Oat
		Tulip	Maples		Okra
			(silver,		Orchard
			sugar)		grass
			Marigold		Peach
			Mint		Pear
			Mimosa		Pecan
			Musk-		Petunia
			melon		Pines
			Oaks		(Austrian,
			(gambel,		jack,
			white)		loblolly,
			Oat		ponderosa,
			Parsley		Virginia,
			Parsnip		white)
			Pea		Poplar
			Peach		Pumpkin
			Peanut		Raspberry
			Petunia		Rhubarb
			Pines		Rose
			(ponderosa,		Soybean
			Scotch,		Spinach
			white)		Squash
			Potatoes		Strawberry
			(sweet,		Sumac
			white)		Sunflower
			Pumpkin		Sweet pea
			Radish		Sweet
			Salvia		potato
			Scallion		Swiss
			Spinach		chard
			Squash		Tomato
			Straw-		Tulip tree
			berry		Turnip
			Swiss		Verbena
			chard		Violet
			Sycamore		Wheat
			Tobacco		Zinnia
			Tulip tree		
			Turnip		
			Verbena		
			Walnut		
			Wheat		
			Weeping		
			willow		

B. Tolerant Plants

PAN	HCl	HF	O₃	C₂H₄	SO₂
Azalea	Arborvitae	Ash	Coleus	Beet	Arbor-
Begonia	Begonia	Cattleya	Cotton	Cabbage	vitae
Cabbage	Grasses	Cherry	Cucumber	Chrysan-	Box elder
Corn	(Kentucky	(flowering)	Euonymus	themum	Castor
Cotton	blue, rye)	Cotton	Geranium	Clover	bean
Cucumber	Hemlock	Chrysan-	Gladiolus	Corn	Chrysan-
Grasses	Ivy	themum	Juniper	Gardenia	themum
(Kentucky	Maple	Cotton	Poinsettia	Gladiolus	Citrus
blue, rye)	(Japanese)	Croton	Yew	Lilac	Corn
Onion	Oak	Elderberry		Lily of	Cucumber
Pansy	(Red)	Gloxinia		the	Gingko
Radish	Soybean	Juniper		valley	Gladiolus
Sorghum	Yew	Orchid		Marigold	Honey-
Wheat		(cattleya,		Oat	suckle
		dendro-		Onion	Iris
		bium,		Pompon	Lilac
		moth)		Radish	Maple
		Schefflera		Sorghum	Oak
		Squash			Onion
		Tomato			Sorghum
		Virginia			Tulip
		creeper			Willow
		Willow			
		Wheat			

Source: Mastalerz, J. W. 1977. *The Greenhouse Environment.* John Wiley and Sons, New York. Reid, M. S. 1992. Ethylene in postharvest technology. In *Postharvest Technology of Horticultural Crops.* A. A. Kader (ed.). University of California Division of Agriculture and Natural Resources Publication 3311, Oakland, CA 94608-1239. Skelly, J. M. and R. C. Lambe. 1974. *Diagnosis of Air Pollution Injury to Plants.* Virginia Polytechnic Institute and State University Extension Division Publication 568, Blacksburg, VA 24061. Woltz, S. S. and W. E. Waters. 1978. Airborne fluoride effects on some flowering and landscape plants. *HortSci*: 13(4) 430-431.

cause damage to sensitive plants. (See Table 15.1.) Some of the more sensitive plants can be injured with concentrations of 0.1-0.2 parts per billion (ppb). Common injury symptoms are necrosis of broad-leaf margins and a tipburn of grass leaves and conifer needles. A yellow mottling develops in corn leaves prior to the appearance of the "burn" and small holes (shotholes) may develop in the leaves of stone fruits.

Ozone

Ozone results from the combustion of petroleum or coal, but sunlight is necessary for its production. Concentrations of a few ppm can cause injury to sensitive plants. (See Table 15.1.) Injury is often associated with smog damage. Combinations of O_3 and sulfur dioxide can cause damage before it would be observed from the presence of either one alone. Symptoms of O_3 damage include tissue collapse, necrosis, stiple markings, flecking, chlorosis, bronzing, stunted growth, depressed flowering, and leaf drop.

Sulfur Dioxide

Sulfur dioxide (SO_2) is given off by smokestacks of industrial plants, chemical factories, and power plants where coal or oil containing sulfur is burned. The SO_2 can be a source of S for plants when brought down as rain, but relatively small quantities (as little as 0.3-0.5 ppm) of the gas can cause injury to sensitive plants. (See Table 15.1.) Symptoms of damage are yellowing and gradual bleaching, with dry, white to straw-colored marginal or interveinal blotches in broad leaves, and light-tan to white streaks on either side of the midvein in grass leaves. The symptoms in conifers are reddish-brown necrosis with adjacent chlorosis of the needle tips. Much of the damage attributable to "acid rain" probably is due to SO_2, as it combines with water to produce sulfurous acid (H_2SO_3). The use of tall smokestacks permits the gas to be carried great distances by prevailing winds before it returns to the earth.

Ethylene

Some ethylene is present in the atmosphere, but its harmful effects derive primarily from a buildup of concentrations in enclosed structures and so will be covered extensively later in the chapter.

Measuring Air Pollutants

Most air pollutants can be measured by on-spot color tubes or gas chromatography.

The on-spot color tubes are the simplest and most rapid, but sensitivity may not be great enough to pick up the very low concentrations which can be harmful to sensitive plants. A wide variety of color detector tubes can be obtained from SKC Inc., 334 Valley View Road, Eighty Four, PA 15330-9614. To activate tubes, a known volume of air is drawn through the glass detector tube by means of a hand bellows. The quantitative concentration of the gas is revealed by the length of the color band produced by the pollutant. Tubes available to test for pollutants and the effective range in ppm for each are as follows: ammonia (2-30), butane (100-800), chlorine (0.3-5), fluorine (0.05-8), methyl bromide (3-100), natural gas (qualitative), mercury vapor (0.1-2*), nitrogen dioxide (0.5-25), nitrous fumes of NO + NO_2 (0.5-10), ozone (0.05-1.4), propane (5000-13,000), and sulfur dioxide (0.1-3).

Accurate measurement of very low levels of most pollutants can be made by gas chromatography, but instruments cost in the range of $7000-10,000. Although it requires more time, it is possible to remove air from rooms or containers by means of glass syringes or sampling tubes and forward them to laboratories equipped with a chromatograph. A basic service providing ethylene analysis of four air samples for a cost of $35 is provided by Prince and Prince, Columbus, OH 43218.

Protection from Air Pollutants

Damage from air pollutants can be reduced by (1) the use of plants that are resistant to the pollutant, or (2) the use of protective sprays. Ultimate control of the problem lies in greatly reducing these toxic materials in the air.

Resistant Plants

Crops that do comparatively well in the presence of the various pollutants are given in Table 15.1. Cultivars of different plants vary as to their resistance to these atmospheric pollutants.

---

*weight/volume

Protective Sprays

Sprays of ascorbic acid, ozoban (a mixture of ascorbic acid and other ingredients that yields potassium ascorbate when mixed with water), or the fungicides Ferbam, Maneb, Thiram, and Zineb have given some protection against pollutants.

POLLUTANTS IN ENCLOSED STRUCTURES

Plants in enclosed structures are susceptible to general air pollution as discussed above, but they can also be seriously affected by locally produced pollutants. Because of the limited air ventilation, these materials can at times cause serious damage to a number of plants.

There are at least five major causes of damage:

1. Gases used for heating or produced by heaters that are not properly vented,
2. Noxious fumes released from wood preservatives,
3. Accumulation of ethylene,
4. Free ammonia from fertilizers or refrigerating equipment, and
5. Mercury vapors released from mercury paints or mercury-filled thermometers, thermostats, and switches.

Pollutants from Heat Sources

Heating greenhouses often produces several gases which can be harmful to plants, primarily from the incomplete burning of fuel. Often added to these pollutants are varying amounts of propane, butane, or natural gas from leaks or spills.

Although there are gas-leak detectors capable of measuring small quantities (2 ppm) of butane, propane, and natural gas, the measurement of combustion products is more difficult because of the wide variety of such pollution sources. The measurement of ethylene, one of the potential pollutants resulting from incomplete burning, is somewhat easier, and its importance and measurement are considered later in the chapter

Avoiding Problems

The simplest way to handle pollutants from heating gases and from incomplete combustion is to avoid their introduction by carrying out combustion away from growing plants or harvested commodities. Combustion activity includes the use of gasoline and diesel-fired vehicles or equipment, as well as all types of heaters. If this is not possible, some measure of control can be gained by properly venting or exhausting the noxious gases.

Wood Preservatives

Gases given off by some wood preservatives can cause considerable damage to growing plants or harvested commodities. Pentachlorophenol or creosote-treated wood can emit toxic fumes for a period of about five years.

Control

Damage can be avoided by the use of copper arsenate, copper chromate, or copper naphthenate to treat wood, or by the use of treated wood sold as Wolmanized wood. Painting pentachlorophenol-treated wood over twice with B-1-N paint (provided the wood has not been oiled) or sealing it with a two-phase epoxy paint limits the emission of fumes to nontoxic levels.

Ethylene

Ethylene is a naturally produced hormone which aids in the flowering or ripening of fruits. Rates of production are high in germinating seeds, meristematic tissue, flowers (particularly as they fade), ripening fruits, and all decaying vegetation. Production is higher in climacteric fruits (those that produce large quantities of heat at harvest) such as avocado, banana, cherimoya, cantaloupe, and tomato, and in certain flowers, such as carnations and Vanda orchids (which have the highest recorded rate of ethylene production by higher plants).

Ethylene production by the plant is ethylene stimulated as plants are placed under stress or as they mature. Harvested commodities continue to produce ethylene and great amounts are liberated as they decay. The hormone can also be produced by gasoline- and diesel-fired engines and by the burning of manufactured gas.

Benefits from Ethylene

Besides serving useful functions in the plant, ethylene is used by man for shuck loosening to ease the harvesting of pecans and walnuts, for hastening the ripening of several fruits, and for flower and sprout induction of flowering bulbs. Both ethylene gas and its precursor, ethephon, are used for such purposes.

Ethylene in concentrations of 10-100 ppm is used in ripening rooms to hasten the ripening of harvested commodities. The lower concentrations are used in tight rooms or if ethylene is released gradually by trickle systems. Temperatures in the range of 65-77°F, an RH of 90-95 percent, and adequate air circulation and ventilation are necessary to ensure adequate distribution of the ethylene and to prevent an accumulation of CO_2.

Damage from Ethylene

The presence of ethylene is a problem when it hastens maturity beyond acceptable marketing stages or speeds it up so rapidly that it is not possible to produce a marketable product. The gas can become a serious problem for sensitive plants if it is allowed to accumulate in greenhouses, storage areas, packing sheds, or transport vehicles.

Very little ethylene is needed to hasten maturity or shorten the shelf life of a number of commodities. Threshold limits for several fruits are in the area of 0.1 ppm, with maximum rates of ripening taking place at 1 ppm. As little as 10 ppb can affect the opening of roses, and exposure of carnation's flower buds to concentrations of as little as 30 ppb can prevent them from opening ("sleepy flowers"). Such low concentrations are easily surpassed in enclosed areas where the gas can accumulate. Harmful concentrations are often readily reached if the ethylene produced by combustion is added to that which is produced naturally.

Reducing Losses from Excess Ethylene

Losses are reduced by several strategies:

1. Avoid plant stress during growth stages;
2. Remove gas- and diesel-fired vehicles from packing sheds and storage areas;
3. Avoid burning trash or wood near or in growing, packing, and storage areas;
4. Cool harvested commodities as soon as possible;
5. Store and transport harvested commodities at the lowest temperature possible without causing cold-induced stress;
6. Spray ornamentals or pulse cut flowers with silver thiosulfate prior to shipment;
7. Use controlled atmosphere of limited O_2 and increased CO_2;
8. Lower O_2 content of stored commodities by using hypobaric storage;
9. Remove ethylene from storage or shipping containers by proper ventilation; and
10. Use materials such as potassium permanganate, and activated or brominated charcoal in storage rooms to absorb the ethylene, or use ultraviolet lamps or catalytic oxidizers to destroy it.

Measurement of Ethylene

Both exploitation of the beneficial uses of ethylene and the inhibition of its harmful effects are made easier by accurate measurement. Often, temperature and humidity measurements are also needed to maximize results.

A plant bioassay method, as well as gas detector tubes and chromatography, can be used to detect ethylene. The bioassay method consists of placing potted vigorous marigold or tomato plants in a suspected area for a period of 24-48 hours at a temperature of 60°F or higher. The presence of as little as 0.1 ppm of ethylene manifests in the plant as epinasty, or a downward bending of the leaf. Curling and abnormal lengthening of leaf tips may also appear with continued exposure.

Free Ammonia

The presence of very small quantities (less than 1 ppm) of free NH_3 in greenhouses, storage areas, or transport vehicles can be detrimental to growing or harvested crops. The NH_3 may be derived from liquid fertilizers containing free NH_3, from freshly applied manure, and from urea that is surface-applied to soils with large amounts of organic residues or high pH. Free NH_3 also may be derived from ice-making equipment that uses ammonia.

Control

Avoidance or exclusion is the primary means of control. All forms of fertilizers containing free NH_3 must be eliminated from growing areas. In addition, avoid (1) nitrogenous fertilizers, such as urea, that easily give off NH_3 when surface-applied to soils with high OM contents or high pH values; and (2) large amounts of surface-applied manure. While it is best to avoid using urea and manure in high pH soils, they may be used safely if they are incorporated quickly into the soil.

All compressors using NH_3 must to be isolated from production, storage, and transport areas.

Mercury

Mercury vapor, released from mercury paints, spilled mercury from thermometers, greenhouse thermostats, or mercury switches, can retard a number of plants (begonia, camellia, carnation, coleus, fuchsia, gardenia, geranium, hydrangea, Easter lily, calla lily, marigold, rose, salvia, and tomato). As little as 0.01 ppm of mercury vapor, can injure sensitive plants, such as the rose cultivar 'Better Times.' Typical symptoms include retardation of growth, hardened appearance, failure of new shoots to develop, and a drastic reduction of flower production.

Control

The problem can be largely avoided by eliminating the use of mercury paints, and using alcohol rather than mercury thermometers in plant growing areas. Care must be taken in repairing or replacing thermostats to avoid spilling mercury or removing it if it is released.

The release of mercury vapor from paints that may contain a mercury fungicide can be lessened by coating the painted surfaces with a mixture of five parts lime sulfur, ten parts wheat flower, and 100 parts water. If the greenhouse is empty, elemental sulfur can reduce damage from mercury paints, but may discolor white paints.

Measuring Mercury

In addition to gas detector tubes, small amounts of mercury can be measured in laboratories by passing the contaminated air over gold leaf, vaporizing the mercury, condensing the vapors and treating the condensate with iodine to produce the colored mercuric iodide. The method is laborious, time-consuming, and costly. A less time-consuming method of detecting mercury is available by using an instrument that measures the ultraviolet absorption of mercury vapor at 2357 Angstrom units, but its maximum sensitivity of about 0.01 ppm of mercury is not sufficient to detect the low levels capable of injuring very sensitive rose cultivars. Also lacking sufficient sensitivity for this purpose are the rapid Sensidine detector tube (with a sensitivity of 0.05 milligrams of mercury vapor per cubic meter of air), the SKC Color Detector Tube (with a sensitivity of 0.1 milligrams per cubic meter), and the time-consuming bioassay method using pinto beans (which can measure 0.06-0.9 ppm of mercury vapor per cubic meter of air).

AUTOMATIC CONTROLS

Various devices are used to monitor gases in enclosed structures. Instruments capable of monitoring one or more of the following: Cl_2, CO_2, F_2, HF, H_2S, NH_3, NO_2, Oxides of N, and SO_2 are manufactured or distributed by CEA Instruments, Emerson, NJ 07630; Spectrum Technologies, Inc., Plainfield, IL 60544; SKC Inc., Eighty Four, PA 15330; and DeltaTrak, Inc., Pleasanton, CA 94566.

Chapter 16

Wind Speed and Direction

The velocity and direction of wind can affect plants in various ways and they need to be considered in several agricultural applications. As wind velocity increases, plants tend to lose water more rapidly. The loss can result in reduced growth if the plant fails to replenish water fast enough to avoid plasmolysis of cells.

WIND AND COLD-INDUCED STRESS

Wind velocity and direction have an effect on cold-induced stress. Increased velocity reduces the danger of frost forming, but it can increase damage by removing cloud covers that hold heat close to the earth or by dehydrating plants. Damage resulting from winds can be much greater if temperatures drop below the freezing point. Wind speed affects the amount of water that needs to be applied to limit damage from cold and determines whether sprinklers can be used effectively. (See Table 13.5 in Chapter 13.) Shifts that produce a southerly wind (Northern Hemisphere) or that could bring in warmer air as it passes over a large water body can indicate that there is no further need of temporary measures designed to lessen damage from cold stress.

Air Mixing

If winds are less than about 5 miles per hour, cold air (which is heavier than warm air) settles close to the earth below a layer of warmer air. Wind machines, airplanes, and helicopters which can mix the warm air with the colder air are useful in reducing cold damage.

WIND SPEED AND SPRAYING

High winds can reduce the efficiency of applied sprays and by moving materials beyond the target can cause serious problems of phytotoxicity and even threaten human health. By inducing water stress in plants, high winds can be instrumental in causing plant injury from a number of spray materials, particularly as the concentrations of materials in the spray tank are increased. As a general rule, it is wise to discontinue spraying if the wind speed exceeds 10 miles per hour. It may be wise to terminate application at wind speeds over 5 miles per hour if the spray is particularly toxic (2,4-D herbicides) and/or if there is no buffer between the sprayed area and nontarget sensitive plants.

Safe application of sprays can be accomplished in winds up to 12 miles per hour, but special care must be taken. Measures that aid in satisfactory spraying at higher wind velocities are: (1) using the largest spray droplet that is recommended for a particular chemical or one that gives satisfactory coverage; (2) lowering the spray boom to the lowest height that will still give adequate coverage; (3) decreasing the pressure as much as possible while still obtaining adequate coverage; (4) using adjuvents if permitted by label; (5) spraying when the soil is cool, such as in early morning or late afternoon, after rain or irrigation, and 6) spraying only if air temperatures are low and there is no danger of a temperature inversion. The presence of inversions can be detected by noting the movement of smoke from a small fire or smoke bomb. An inversion is indicated if the smoke moves horizontally close to the ground (less than 1000 feet).

MEASURING WIND SPEED

Wind speed can be measured by wind anemometers or wind meters. These can be hand-held types or mounted on stakes or buildings. Some types supply wind direction as well as speed, and some are equipped with printers to supply a permanent record. They can be combined with various other instruments that measure temperature, humidity, rainfall, and air pressure to provide a complete weather station.

Additional Reading

Anonymous. 1983. Light specifications for interior plants. *Nurserymen's Digest* (April): 24-25.

Bartok, J. W., Jr. 1993. Cooling without fans. *Greenhouse Grower* (July): 37-38.

Dennis, F. G. 1984. Flowering. In *Physiological Basis of Crop Growth and Development*. M. B. Tesar (ed.). American Society of Agronomy, Crop Science Society of America, Madison, WI 53711.

Hashimito, Y., G. P. A. Bot, W. Day, H. J. Tantau, and H. Nonami (eds.). 1993. *The Computerized Greenhouse*. Academic Press, Orlando, FL 32887.

Janick, J. 1979. *Horticultural Science*. W. H. Freeman and Company, San Francisco.

Kader, A. A. (ed.). 1992. *Postharvest Technology of Horticultural Crops*. Publication 3311. Division of Agriculture and Natural Resources, University of California, Oakland, CA 94608-1239.

Markey, A. E. 1992. Develop a drift management plan. In *Insect Control Guide*, pp. 13-14. Meister Publishing Co., Willoughby, OH 44094.

Mastalerz, J. W. 1977. *The Greenhouse Environment*. John Wiley and Sons, London, Sydney, Toronto, Santa Barbara.

Oglevee, J. R. 1993. The smart greenhouse. *Greenhouse Grower* (July): 96-98.

SECTION VI:
TROUBLESHOOTING

Despite the best preparation and prevention methods, problems of plant growth serious enough to cause economic losses will occur. The rapid recognition and determination of the cause is essential to limit the amount of damage. Early recognition allows for timely correction, reducing the severity of the problem and making it more manageable. Early corrective measures may also prevent the problem from recurring in later plantings.

Early recognition requires routine examination of the crop by competent personnel. Routine examinations, such as made for IPM (Integrated Pest Management), can often detect problems at very early stages, thereby preventing serious losses. Regardless of the point when the problem is recognized, selecting the proper mode of correction will depend on an accurate diagnosis of the problem.

Chapter 17

Appraising the Problem: Systematic Approach

An accurate appraisal of the cause of a plant problem is hastened by using a systematic approach that will vary depending on the crop and its stage of growth. All plant parts need to be examined and all data pertaining to crop growth need to be available and subject to evaluation. In addition to past data, it is important to examine current data related to soil and plant nutritional status, soil and air temperatures, solar radiation, soil moisture status, presence of soil compaction, presence of pests, and measures used for controlling pests. Ideally these problems are resolved at the farm or in the greenhouse, but often it will be necessary to collect various samples and submit them to a laboratory for additional diagnosis.

Basic procedures of soil and plant examination are common for all plants, but problems and symptoms are often plant-specific. Knowing these procedures can aid in identification of specific problems. A guide for diagnosing corn problems, presented in Table 17.1, can be used as a model for other crops as well. Although other crops may have many different problems, much of the approach in delineating the problem for corn can be used effectively for other crops.

Diagnosing several different types of injury to corn, cotton, small grains, and soybeans is made easier by use of Crop Injury Diagnostic Guides for these crops, obtainable from Agri-Growth Research, Inc., Hollandale, MN 56045.

STAGES OF GROWTH

While complete examinations are needed at all times, the evaluation of certain characteristics will be more rewarding at particular stages of growth.

TABLE 17.1. Corn diagnostic guide

General appearance	Specific symptoms	Cause
BEFORE EMERGENCE (Stages 0-0.2)		
Skips in rows where plants fail to emerge	No seed planted	Planter malfunction; empty planter box; irregular seeding depth
	Seed not sprouted	Seed not viable; nutrient injury
	Seed swollen, but not sprouted	Seed not viable; soil too cold or too wet; anhydrous or aqua ammonia injury
	Rotted seed or seedlings	Fungal seed rots or seedling blights; anhydrous or aqua ammonia injury; wet soil
	Sprouts twisted, leaves expanded underground	Soil crusted, compacted, or cloddy; seeds planted too deep in cold, wet soil; herbicide or soil insecticide injury
Seeds eaten, dug up, or sprout cut off	Seeds hollowed out	Insect damage
	Unemerged seedlings pulled up and seed eaten	Crows, pheasants, blackbirds
	Seed dug up and eaten or unemerged seedlings pulled up and entire plant eaten	Mice, groundhogs, ground squirrels, gophers, skunks
EMERGENCE TO KNEE HIGH (Stages 0.2-2)		
Scattered problem spots of dead or poorly growing plants	Uneven growth	Drainage; soil compaction; variations in planting depth or soil moisture; poor growing conditions (cold, wet, dry, etc.); seed bed not uniform (cloddy)
	Plants stunted, wilted, and/or discolored	Nematodes; seedling wilt; damping-off and seedling blight; Stewart's bacterial wilt; root rot complex
	Sudden death of plants	Frost; lightning
Widespread wilting	Upper leaves roll and appear dull	Drought conditions; insects
Plants discolored	Leaves appear sandblasted; leaves pale green	Wind damage; insect damage

General appearance	Specific symptoms	Cause
	Yellowing between leaf veins followed by death of leaves, beginning from tips	Herbicide damage
	General yellowing of upper leaves	Magnesium deficiency
	General yellowing of lower leaves	Excessive moisture
	Purpling or reddening of leaves from tip starting with lower leaves; tips may turn dark brown and die	Phosphorus deficiency; compacted soil; cold weather; insect damage
	Leaves slowly turn white to tan and die	Herbicide damage
	Irregular light gray or silvery blotches on both sides of leaves on the east side of affected plants	Sunscald (occurs when chilly, dewy nights are followed by clear, sunny mornings)
	Light streaking of leaves developing into broad band of bleached tissue on each side of midribs; midribs and margins remain green; stalks and leaf edges may appear tinted red or brown	Zinc deficiency
	White or yellow stripes between leaf veins	Acidic soil-magnesium deficiency
	Distinct bleached bands across leaf blades; leaf tips may die back; leaf tissue may collapse at discolored bands, resulting in the leaf folding downward at this point	Air pollution damage
Plants discolored and stunted	Leaves yellow; plants spindly and stunted	Nutrient deficiency
	Leaves purple or red, especially leaf margins; stunting	Herbicide injury phosphorus deficiency

TABLE 17.1 (continued)

General appearance	Specific symptoms	Cause
Plants discolored, malformed, and/or stunted	Leaves yellow, plants spindly	Disease damage
	Slight yellow-green tint; severe stunting; leaves do not unfold; leaf tips stick together	Calcium deficiency
	Leaves yellow, not fully expanded; roots sheared off or dried up	Overapplication of anhydrous or aqua ammonia
Plants stunted and/or malformed	Leaves fail to unfurl properly; often leafing out underground, plants may be bent, lying flat on soil surface	Herbicide damage
	Leaves stunted, twisted, and may appear knotted; shoots and roots malformed; general stunting	Herbicide damage
	Plants bent or twisted; irregular rows of holes in unfolded leaves	Insect damage
Lesions on leaves	Spots of dead tissue on leaves	Herbicide injury; disease damage
Plant tissue removed	Plant cut off at ground level; chunks of leaf tissue removed	Insect damage
	Ragged holes in leaves	Hail damage (all plant material in area damaged but may be more severe on one side of plants); insect damage
	Leaves shredded or torn	Wind damage
	Rows of circular to elliptical holes across leaves	Insect damage
KNEE HIGH TO TASSELING (Stages 2-4)		
Severe wilting and/or death of plants	Sudden death of plants	Lightning (all plant material in an approximate circle suddenly killed; plants along edges may be severely to slightly injured; severely injured plants may die later)

General appearance	Specific symptoms	Cause
	Dieback of leaves; wilting, then drying up of leaf tissue beginning at tips	Molybdenum deficiency; air pollution injury
	Plant tissue grays and dies (similar to frost damage)	Disease damage
Plants discolored	Plants yellowing, beginning with lower leaves	Lacks nutrients; drought or ponding produces nitrogen deficiency
	Plants yellowing, beginning with upper leaves	Copper deficiency (stalks soft and flexible; stunting of plants); high temperature (leaves yellow, then roll and bleach; associated with low soil moisture and hot, dry winds)
	Leaf margins yellow, beginning at tips; affected tissue turns brown and dies	Potassium deficiency
	Purpling or reddening of leaves from tip backward affects lower leaves first; leaf tips may turn dark brown and die	Phosphorus deficiency
	Yellow to white interveinal striping of leaves	Disease injury; magnesium deficiency (yellow to white striping usually developing on lower leaves; red-purple along edges and tips; stunting possible); boron deficiency (initially white, uneven spots develop between veins, may coalesce to form white stripes that appear waxy and raised; plants may be stunted)
	Pale green to white stripes between leaf veins, usually on upper leaves	Iron deficiency

General appearance	Specific symptoms	Cause
	Pale green to yellow interveinal chlorosis of upper leaves; lower leaves olive green and somewhat streaked; severe damage appears as elongated white streaks with centers that turn brown and fall out	Manganese deficiency
Plants malformed or discolored	Plants lodge or grow up in a curved "sledrunner" or gooseneck shape preventing normal brace root development	Insect damage; herbicide injury; root disease complex; mechanical injury; hot, dry weather and winds
	Stalk breakage	Disease or insect damage; herbicide injury
Lesions on plants	Tan, oval to circular, usually with concentric zones on leaves	Helminthosporium; leaf spot
	Pale green or wavy leaf streaks (irregular or wavy-margined in shape); streaked areas die and become straw-colored; veinal leaves, especially of sweet corn, may die	Disease damage, usually transmitted by corn flea beetles; air pollution
Plant tissue removed	Ragged holes in leaf	Hail damage (all plants in area affected; damage may be more severe on one side of plants)
	Shredding, tearing of leaves	Wind damage
	Longitudinal slits between leaf veins	Insect damage
	Green upper layer stripped from leaves	Insect damage
	Holes bored in stalks	Insect damage

Source: 1992. Field diagnostic kits proliferate. *Solutions* (March/April), 36:(3) 30, 31. Reprinted by permission of the Agricultural Retailers Association, St. Louis, MO 63146.

Early Season

The vigor of young seedlings can be helpful in predicting later growth and production. The stand and early growth are indicative of seed quality, extent of disease and insect damage, suitability of land preparation and depth of planting, and sufficiency of moisture and nutrients. This is a good time to look for deficiencies of P and Zn in plants, as these shortages will be apparent very early.

Placement of banded fertilizer often is discernable for a few weeks after planting. Improper placement can be responsible for weak growth (bands too far from emerging seedlings) or injury (bands too close).

Examination of the tap root of young seedlings can reveal possible fertilizer damage, soil compaction, and several root diseases.

Midseason

This is a good time to look for off-colors, abnormal plants, and insect and disease damage, and to check the plant's leaf size, leaf and stem tips, and extent of root systems from a standpoint of nutrition. Careful observations as to the presence and extent of wilting, along with careful attention to soil moisture conditions as revealed by tensiometers, evapotranspiration, and soil examination, can help to reveal a number of problems at this stage. Clues to problems are also forthcoming from soil and plant analyses, although they may have to be repeated within a few weeks to fully reveal the extent of deficiency or excess. Monitoring for insect and disease attacks during this period also helps explain various problems. Evaluation of bee activity, weather conditions for pollination, lack of set or fruit drop, and poor tillering or grain filling in cereals may explain many problems of crop production.

Late Season

Lodging of grains, stalk diseases, and poor color, poor shape, or damage to fruits reveal a number of problems that influence the yield or quality of many crops.

TOOLS

As with so many enterprises, the use of tools can be of great assistance in troubleshooting. The specific tools to use will vary with the crop, the nature of the problem, and whether the crop is grown in the field or in a protected structure.

These are some of the more useful tools:

- spade or shovel
- trowel
- Hoffer soil tube or other means of taking soil samples
- knife
- clean bucket
- sample bags
- hand lens of 10-20 power
- a sweep net to catch insects
- conductivity meter (portable or pen type)
- pH meter (portable or pen type)
- test kits for evaluating nutrient deficiencies
- test papers for NH_4-N and NO_3-N
- ion electrodes for NO_3-N and K
- soil thermometer to evaluate poor stands of early crops
- an infrared thermometer to evaluate plants under stress
- field map
- camera
- sturdy notebook

In addition to most of the above, the following can be useful in evaluating problems in protected structures:

- light meter
- wet- and dry-bulb thermometers
- gas-detecting tubes

CROP HISTORY

Finding a solution to plant problems is facilitated by obtaining a full history of the crop. Facts concerning the current crop that can be useful are:

1. soil preparation;
2. preplanting treatments of manure, lime, fertilizer, nematicides (amounts and manner in which they were applied);
3. date of planting;
4. weather conditions at planting and following planting;
5. rainfall and/or irrigation levels;
6. amounts, analysis, and position placement of fertilizer side-dressings;
7. insect and disease control;
8. how and when sprays were applied;
9. soil and plant analyses; and
10. onset of the problem and what treatments have been applied to date in an effort to correct it.

Examination of preparation, fertilizer, and spray equipment can be very helpful, especially if they can be viewed in actual operation. Also useful are recorded temperature, humidity, and CO_2 content of enclosed structures and transportation vehicles.

Histories of previous crops that supply information as to rotation of crops, fertilization and spray programs, and presence of previous problems can also be enlightening.

PLANTS

Careful examination of the affected plant(s) is necessary for effective diagnosis. Examination of a nearby plant or plants not affected or only marginally affected will help in making a proper diagnosis. Both types of plants need to be examined from the standpoints of (1) overall plant location, (2) plant stress, and (3) general plant appearance.

Overall Plant Location

Plants may be doing poorly because of their location in a field. If affected plants are in a low spot, excess water may be the cause. Low spots are appreciably cooler during cold, still nights, and the lowered temperatures during nights approaching the freezing point could do appreciably more damage to plants in low areas. Plants particularly susceptible to such damage are warm weather crops and a number of crops in bloom. The presence of poor plants only in high spots can signify lack of moisture.

Poor plants in a circular or elongated patch may signify nematode problems. The chances of nematode problems are increased if the area of damage increases each year and is increasing in the direction of equipment movement. Turf damage in a circular pattern could be due to disease. Sudden death of plants in an extended patch or circular area could be the result of a lightning strike. Poor growth in circular patches in fields with overhead irrigation could be the result of malfunctioning nozzles.

Poor plant growth in a row or part of a row often results from poor distribution of fertilizer or other soil amendments. In drip-irrigated fields, clogged tubes could prevent the satisfactory flow of water or water plus nutrients.

Poor plants at the edges of fields may be the result of poor application of fertilizer, manure, lime, or other amendments. Sometimes poor plants are the result of doubling up materials as applications overlap, or of compacted soils if equipment made many passes over the same area. Poor plants at the edges of fields can also be the result of migrating pests—insects, birds, and diseases.

In protected structures, small changes in position may mean appreciable differences in the amounts of light, water, fertilizer, or temperature to which the plant is exposed. During the summer months, proximity to cooling devices can make an appreciable beneficial difference. In the winter, there may be not only appreciable differences in critical temperature, but to exposure of various pollutants as well. In northern latitudes, the positioning of plants during the winter months so that they receive less light (shade from lighting fixtures, heating equipment, or other plants) can have serious negative effects because of borderline light levels overall.

Plant Stress

Many plant problems are initiated by putting plants under stress. The stress may be due to various factors:

1. temperatures too high or too low;
2. excesses or shortages of water;

3. insufficient light intensity or inappropriate day length (more common for plants grown in greenhouses);
4. poor soil aeration resulting from compacted soils or high water tables;
5. deficiencies or excesses of plant nutrients;
6. high or low soil pH values;
7. poor placement of fertilizer;
8. excessive soil or water salt levels;
9. heavy pest infestations;
10. damage resulting from the wrong pesticide or the correct one being applied too frequently or from carryover of a pesticide;
11. excessive levels of salts on leaves as a result of foliar applications of nutrients, insecticides, fungicides, or overhead application of irrigation waters high in Cl, Na, or bicarbonates; and
12. excess applications either to soil or leaf of B or F from irrigation water or fertilizer.

The stress may result in poor growth, in various leaf, stem, and root symptoms, and in greater susceptibility to disease and insect attack. Some of the common symptoms of stress affecting leaf, stem, and root symptoms are noted below. But it is important to note here that many of the effects of stress are indirect as they affect the incidence of disease. Generally, stress conditions favor organisms that gain a foothold on old or dying tissue, such as gray mold caused by *Botrytis cinerea*. Typical examples of several diseases brought on by stress are given below.

Cold, wet, compacted soils favor losses from *Pythium* species. Damping-off seedlings caused by *Pythium* occurs most frequently in poorly drained, unpasteurized soil in trays placed on the ground, but it is seldom a problem if trays are placed on well-ventilated warm benches and the soil is watered with warm water. The severity and frequency of root and stem diseases caused by *Fusarium*, *Pythium*, and *Phytophthoa* are less of a problem if field soils are well aerated and drained. Low light conditions can reduce resistance to *Verticillum wilt* in tomato and can increase the virulence of several wilt pathogens, such as *Fusarium*. Excessive soil moisture, especially combined with high RH, encourages gummy stem blight and gray mold in cucumbers. But a few disease-causing organisms, such as *Rhizoctonia* and *Thielaviopsis*, are favored by dry soils (less than 40 percent MHC).

General Plant Appearance

Poor, thin stands may be the result of poor seed; poor seed storage; old seeds; poor seed bed preparation; planting too deeply; planting in soils that are too hot, too cold, too wet or, too dry; failure to firm soil or exerting excess compaction of seed at planting; damage from birds, rodents, insects, fungi (damping-off), and nematodes; excess salts; poor fertilizer placement in relationship to the seed; or lack of moisture during the germination process.

Unthrifty plants of poor size may be the result of poor seed, nematode attack, disease or insect infestation, poor nutrition, or poor soil water. Stunting or collapse of the plant may be due to a viral problem. Occasionally, poor plants are found in spots with compacted layers, hardpans, or rocky ledges.

Seedlings that turn white, silver, yellow, or purple probably have been exposed to unduly cold temperatures. Those that fall over or collapse from decayed or water-soaked stems can be suffering from damping-off, a fungal disease.

ROOTS

Symptoms on roots can help to diagnose a number of problems. Carefully lifting annual plants or small perennials with a trowel or spade will allow for good observation of the roots. The soil needs to be shaken or washed gently from the roots. Generally, it will be impossible to examine the entire root system of large trees but a worthwhile inspection of the roots can be made by digging narrow trenches 2-3 feet in depth close to the drip line. Roots need to be split open to examine the conducting tissue.

Healthy roots are white to light gray or tan in color and have good numbers of root hairs. There should be little or no bending, crimping, or swelling. Conducting tissue exposed as the roots are split should show no signs of discoloration or plugging.

Infectious disease is indicated by dieback, discolored, dead, or rotted roots, and large galls. If affected by disease, upper parts of the root may show discolored conducting tissue when cut longitudinally.

Insect damage can appear as abbreviated root systems with little or no feeder roots, considerable branching (sometimes in a rosette-type formation) and, dying or dead roots. The larvae of several insects, such as striped cucumber beetle or Japanese beetle, tend to feed on roots, damaging the root system and in a few cases transmitting infectious diseases. Mealy bugs may also cause damage to the roots.

Nutritional problems can be suspected by the presence of short, excessively branched roots that are discolored or that die prematurely. The particular shade of discoloration provides clues to nutritional problems: a brown tip indicates a shortage of Ca; a light brown tip indicates a shortage of P or excess of N; and a dark brown tip indicates excessive K or Cu. Roots that are retarded may be suffering from low P, Ca, or Fe, or excess Al. Those that die prematurely may lack B. Excessively long, thin roots may be the result of low P and N. If thin and light brown, Ca and Mg may be present in excess. Stubby roots can result from low Ca or Zn, or excess Mg or Cu. Poor nodulation of legume roots often results from low Ca levels. An extensive root system can signify shortages of N, especially if the root system is supporting a diminished top with small, thin, light-green leaves.

Excess Al, usually associated with low pH, will tend to produce poor, abbreviated, thickened root systems that are tan in color, much as those that result from Ca and P deficiency. They also may be clubbed (short and thickened), kinked (unduly twisted) with few or no laterals, and breakdown exhibiting blackened and mucilagenous areas.

Nematode damage is indicated by stunted roots with few fine roots; blunt and swollen root tips; many stunted lateral roots near main root tips; roots with necrotic lesions; rotten or dead roots, and root galls. The root galls caused by root-knot nematodes may be confused with other swellings caused by ectoparasitic nematodes or nitrogen-fixing bacteria. The root-knot swelling may appear as small, beadlike individual formations or as massive accumulations of distorted tissue. They are distinguishable from the nitrogen nodules which are easily detached from the roots and usually have a milky pink to brown liquid inside the nodule.

Insufficient oxygen and other stresses of *compact soils* can be suspected by the presence of shallow root systems with darkened roots that tend to branch excessively. If compaction has been associated with flooding, many of the roots may be dead or rotting.

STEMS

The appearance of the stem can be helpful in diagnosing some problems. Stems should be split open so that the conducting tissue can be examined.

Infectious disease is indicated if stems have dieback, cankers, lesions, dark, water-soaked areas near the soil line, gum exudation, breaking away of bark, rots and the presence of discolored conducting tissue, and stems that tend to collapse. Twigs may show galls, cankers, lesions, and dieback.

Insect damage may be manifested as holes in stems (borers), punctured stems (rhubarb curculio), damage to young plants at the soil line (cutworms), or as green, tan, or brown shieldlike structures (scales).

Nutritional disorders are reflected as stem lesions (excess ammonium-N); small stems (shortages of N, P, K, S, Fe, or Cu); spindly and soft stems (low S or excess N); brittle stems (low Ca or Mg); stems with internal discolored tissue (low Ca); hollow stems (low B); spindly stems (low N, P, or Cl). Lack of K results in a shortening of stem internodes giving a telescoping effect to the stems of broad bean, cereals, mustard, potato, and tomato. Reduced tillering in cereals can be the result of N or P shortages. Withered tips with the dieback of several plants is caused by a lack of Cu. Roughening of the bark accompanied by blisters from which gum may exude ("exanthema") is also a typical symptom of Cu deficiency. Multiple branching and severe rosetting of both terminal and lateral shoots of several woody species are symptoms of Zn deficiency. Stem streak necrosis and internal bark necrosis are symptoms of excess Mn.

Insufficient light can result in long, thin stems with a stretched appearance.

Waterlogging can lead to bark peeling from stems. Stems may be soft and/or water soaked and prone to rotting. *Freeze damage* can also cause the bark to peel. Split stems and/or stems that exudate gums are common symptoms of freeze damage.

LEAVES

Careful attention to leaf size, abnormal shape, color, and damage are major keys to solving problems. Differences in the appearance between younger and older leaves can aid in identifying nutritional disorders. The location of the leaves with problems can help identify problems associated with spraying.

Infectious disease is indicated by leaves with marginal burning; premature yellowing or reddening; extensive premature leaf drop; yellow halos; shotholes; pustules; lesions of various types; dark or water-soaked areas turning brown; sootlike, crusty fruiting structures; powdery growth on the underside with curled and distorted leaves; or rust.

Viral diseases are indicated by the mottling, distortion, and stunting of leaves. Distortion, similar to 2, 4-D damage, can result from viral infections on several plants. Ringspot, distorted, and mottled rings also may be present.

Defoliation results from age, but it can be hastened by lack of light and overwatering. Occurring with tropical plants (but not confined to them) is leaf defoliation caused by chilling temperatures, strong drafts (especially of cool air), and wetting plants with cold water.

Wilting can result from either a lack of or an excess of water, or from poor uptake of water due to excessively cold soils during periods of good solar radiation. Poor uptake may also be the result of excess soil salts, or damage to roots and stems caused by several kinds of bacteria, fungi, and nematodes.

Chlorosis is often caused by nutrient deficiencies or excesses, but can also result from lack of light, chilling temperatures, exposure to excess ethylene, and root damage from overwatering at cool temperatures or nematode attack. Chlorotic leaves of lilies, appearing as a band near the leaf tips, can result from sprays of Avid insecticide.

Insect damage is commonly indicated by chewed or tattered leaves, often on edges, but may also be visible as small holes with a shotgun (flea beetles) or lacework (Mexican bean beetle) appearance; distorted shape (leafhoppers); or cupped leaves (mites or aphids). Damage caused by leaf miners will be visible as small tunnels between the upper and lower leaf surfaces. Mature scale insects, usually attached to the underside of the leaf, appear as small protuberances of wax. Mealy bugs produce the appearance of cotton. Thrip damage appears as small white specks on silvery grey leaves. At times dark necrotic areas which may be red or brown can be present. Leaves may become curled or distorted. Leafhoppers tend to produce leaves with a whitened or mottled appearance that may turn brown, yellow, or red. Leaves may also have withered tips and die as if scorched by heat. Leafroller larvae fold or roll leaves while they feed on the inside of this cover. The lace bug can turn eggplant leaves yellow or brown. Whitefly damage can manifest itself as a white sheen on squash and cole crops.

Although mites are not insects, *mite damage* is often associated with that of insects. In addition to the cupping of leaves noted above, mite damage is also apparent as yellow or white specks, mainly on the underside of the leaf, which tend to give the plant a bleached appearance. In severe attacks of spider mites, a web is apparent, and leaf chlorosis with severe leaf drop may result.

Viral damage is indicated by thin, distorted, mottled leaves, and associated with the aphids or whiteflies that transmit the virus. Often, symptoms may be more severe at the edges of a field, corresponding to the initial point of entry by the transmitter.

Nematode damage, which is usually present in the roots, is not commonly manifested as distinguishable leaf symptoms. Often, the only symptoms may take on the appearance of one or more nutritional deficiencies or small unthrifty leaves. A more identifiable symptom is the tendency of the leaves to wilt, with poor recovery following a rain or irrigation. The foliar or spring crimp nematode, living on the leaves of several floral crops, produces brown lesions on older leaves, which may cause premature death of the leaf.

Herbicide damage is reflected by a delay of the first true leaf, and various types of chlorosis, yellowing, or death of leaves. Damage from simazine can reveal itself as bright yellow areas between the veins that remain dark green. Early diuron damage will tend to produce yellow veins in leaves; advanced symptoms are marked by the yellowing of the entire leaf, which is also quite similar to simazine damage. Both herbicides can cause "burning" and death of the leaves in later stages. The uracil compounds, isocil, bromocil, and norflurazone cause chlorosis of the midveins and even the veinlets, producing a lacy pattern. Norflurazone damage results in yellowing, or a pinkish or reddish appearance of the petioles. The leaves of plants affected by chlorophenoxy herbicide damage (2, 4-D and related compounds) tend to be small and twisted, with the tips greatly elongated and showing downward bending (epinasty).

Spray damage can be reflected as a "burning" or hardening of the leaf. Several materials will cause chlorosis. Damage may be limited to a given set of leaves that were sprayed, while newer leaves are normal. If successive sprays have caused damage, it is possible at times to find successive layers of damage.

Spray damage can be caused by excessive concentrations of relatively benign substances and/or specific toxic effects of ingredients. Damage from high concentration of sprays (osmotic effects) will tend to appear quite similar to salt effects, with "burning" of tender foliage, hardening of older leaves, and leaf drop. Damage from toxic effects tends to produce chlorosis and distorted leaves.

Insufficient light tends to produce leaves that are thin, small, pale and have elongated petioles. Normal variegation, which enhances the appearance of many decorative plants, can be absent. The lower leaves of some plants may be chlorotic and there is a tendency for leaves to drop prematurely. *Philodendron oxycardium* placed under conditions of insufficient light will produce successively smaller leaves and will ultimately cease growing.

Excessive light can cause shade plants to produce light green leaves that are unattractive. There is less of a contrast between the dark and light portions of variegated leaves.

Insufficient water will yield leaves that lack luster or may be wilted, and tend to drop prematurely. Excessive water may also yield wilted leaves that tend to drop prematurely, but roots of these plants will tend to be brown and subject to decay.

Nutritional disorders are often manifested as differences in size and color, but these vary with the nutrient, the crop, and the age of the leaf. Leaf abnormalities may be the result of deficiencies or excesses of nutrients. Illustrations or photographs of a number of nutritional disorders are presented in several books and monographs listed under Additional Reading at the end of this section, but some specifics listed below may be helpful in diagnosing some problems.

Symptoms of Nutritional Disorders

Generalized symptoms of nutrient deficiencies are the appearance of pale, small, yellowish-green leaves that may turn yellow or have yellow spots or streaks between the veins. They may be distorted or fail to fully develop and they tend to fall prematurely, often developing "burnt" edges prior to falling. Older leaves have the tendency to develop tints of yellow, and even purple. Deficiencies of Ca, S, B, Cu, Fe, Mn, and Zn first appear on younger leaves while deficiencies of N, K, and Mg are first apparent on the older leaves.

Excesses or toxicities of nutrients tend to manifest themselves as "burnt" edges on older leaves. Such leaves may shed prematurely. In the cases of N and S, the leaves may be deep green. Excesses of P, K, Cu, and Fe are often manifested as deficiencies of other elements which are discussed individually below.

Iron deficiency can be induced by excesses of the heavy metals cadmium (Cd), chromium (Cr), cobalt (Co), nickel (Ni) and vanadium (V) as well as by excess Cu, Mn, and Zn. Iron deficiencies exhibited by plants grown on soil treated with large amounts of sewage sludge are usually due to excesses of heavy metals found in some sludges.

Symptoms of Shortages or Excesses of Individual Elements

A description of symptoms caused by shortages or excesses of the individual elements can be helpful in a preliminary diagnosis of nutritional disorders, especially if they are combined with rapid leaf analysis. Such information can be useful in guiding foliar applications aimed at correcting the problem, but no attempt to apply amendments to the soil should be made without also obtaining a rather complete soil test supplying pH, conductivity, and available nutrients.

Nitrogen

Lack of N results in small thin leaves. The angle between petiole and stem is greatly reduced in many plants, especially barley, oat, wheat, flax, potato, and tomato. Leaves can be light green in color, but often have purple, red, or orange coloring. Leaf bases of cereals may be red-purple; upper leaf surfaces of cauliflower may have orange or red flushes; and purple tinting may be present in the veins of the lower leaf surfaces and petioles of tomatoes. Leaf loss from early dehiscence and senescence is accelerated in many plants.

Excess N is often indicated by highly vegetative plants with large, deep green leaves. Excess N in the soil, especially if combined with high K levels, can lead to excess salts, producing the typical symptoms of wilted or "burnt" leaves.

Phosphorus

Symptoms of P deficiency are quite similar to those of N deficiency, with reduction of leaf size and the production of colors. Deep purple tints are common in corn leaves; red and purple in barley and some brassicas; and bronze tints in older hop leaves, French beans, and red clover.

An excess of P can lead to deficiencies of Cu, Fe, Mn, and Zn. The induction of Fe and Mn deficiencies with excess P do not appear to lead to serious crop losses, but induced Cu deficiency has caused serious losses in citrus. Losses from Zn deficiency related to excess P have been much more common and have produced much greater economic loss.

Large amounts of P, unless accompanied by a good supply of N, can lead to symptoms of N deficiencies.

Potassium

Leaf scorch is a common leaf symptom resulting from K deficiency, and is often preceded by interveinal and marginal chlorosis. The chlorosis usually occurs first in the oldest leaves. Spotting or breakdown of leaf tissue may occur in alfalfa, clover, potato, and tomato. Black currant and subterranean clover leaves may have pronounced red or violet spots before scorch is apparent. Barley leaves may have bleached necrotic lesions. Rosetting of leaves is common in beet, carrot, and parsnips.

Overuse of K can lead to symptoms of N, Ca, and/or Mg deficiency. As noted above, excess K can produce symptoms of excess salts.

Sulfur

Lack of S tends to produce leaf symptoms that with respect to leaf size and color are quite similar to those of N shortages. The symptoms are dissimilar in the fact that some symptoms–at least for cocoa, cotton, pecan, soybean, sugar cane, and tobacco–tend to appear first on the younger rather than the older leaves.

Excess S can lead to leaf symptoms similar to those produced by inadequate N. An oversupply of sulfate-S can contribute to excess salts and corresponding leaf symptoms.

Calcium

Lack of Ca tends to affect growing points and developing leaves. Emerging leaves of cereals may remain trapped and fail to unfurl. Those that develop tend to be rolled and chlorotic. The leaves of the spinach beet may break down and turn black. "Tip burn" or the death of young leaf tips is common in lettuce, pea, and strawberry. The tips and leaf margins of young rubber leaves can be bleached and scorched.

An oversupply of Ca is manifested in leaves as shortages of B, Mg, or K.

Magnesium

Deficient Mg is often revealed as interveinal chlorosis, appearing first on older leaves. A green margin may persist in the leaves of apple, sweet cherry, pea, rubber, potato, and red, white, and subterranean clovers. This margin may turn yellow or develop tints of brilliant orange, red, or purple. Tints may also be present in chlorotic leaves of black currant, broccoli, cauliflower, sweet cherry, subterranean clover, cotton, gooseberry, and turnip. A combination of orange and pale green areas along interveinal areas is commonly found in oat leaves.

An oversupply of Mg can lead to symptoms of Ca or K shortages.

Boron

Shortages of B appear as a malformation of leaf veins accompanied by chlorosis, necrosis, or inhibited growth of the basal margins of young leaves. Very young leaves tend to be bleached, shriveled, or blackened. The youngest leaves of cereals may fail to uncoil and remain trapped in leaf bases. Developed leaves often are thicker, very turgid, and brittle. The laminas of grape and raspberry leaves may tear. Petioles of tomato leaves tend to collapse, while those of celery can develop transverse cracks.

Leaves affected by excess B tend to show progressive marginal necrosis, with pronounced curling. Necrotic spots along the main veins and body of the leaf and orange-brown or reddish-brown lesions along the margins of older leaves are common. In citrus, leaf tips tend to be yellow, the leaf tips and margins may be "burnt," and brownish resinous spots can be present on the undersides of the leaves. Chlorosis and/or premature shedding occur in several plants. Young leaves of poinsettia are small and chlorotic, with a tendency to curl; older leaves are glossy and petioles can be dark purple. Prune leaves are coarse, have thickened midribs, and tissue near the midvein is bronzed. Small irregular necrotic spots near the midrib can fall out. Margins of leaves may roll upward and when present near the growing tip become chlorotic and blackened.

Chlorine

Although excesses of Cl as chlorides are much more common than are shortages, lack of the element can cause crop losses. Wilting is a common symptom of deficiency but can be confused with that caused by excess chlorides and other accumulated salts. Chlorosis is also a common symptom. Leaves may also be narrow, become cupped, have prominent raised veins, and develop bronze pigments.

Symptoms of Cl toxicity include burning on leaf tips or margins, bronzing, premature senescence, and infrequent chlorosis.

Copper

Lack of Cu results in varied symptoms in different plants. Generally, new leaves are affected and appear stunted, deformed, and chlorotic. Some leaves may be blue-green. The deficiency seriously rolls or curls the leaves of many plants, especially those of cereals, flax, and tomato. Emerging oat and wheat leaves are white and tightly rolled. Sometimes the rolled leaves are coiled in a spiral that can reverse direction along its length and may have a white tip, characteristic of "reclamation disease." Green pepper and tomato leaves may have interveinal crinkling and marginal wilting. The spiral-twisted needles of Sitka spruce often have necrotic lesions.

An excess of soil Cu appears as an Fe deficiency in many plants. Plants are usually stunted with limited foliage. High concentration of Cu sprays can cause necrotic, elliptical lesions, some of which may be surrounded by chlorotic halos.

Fluorine

No deficiencies of F have been noted in the field, but toxicities from excess have affected a number of plants. Some sensitive plants are "Baby Doll" plant, calatheas, dracaenas, gladiolus, lilies, marantas, and "Ti plant." Chlorosis and subsequent necrosis of the tips and margins of elongating leaves are common symptoms. Spotting near the leaf tips is often followed by scorching, which may affect major portions of the leaf tips.

Iron

A deficiency of Fe is almost always indicated by an interveinal chlorosis of young, rapidly expanding leaves. Chlorosis may appear as alternate bands or affect the entire leaf of cereals and grasses. Completely white leaves may not collapse for a while, but it is not uncommon for chlorotic cereal leaves to have bleached or brown lesions develop in the interveinal areas and for the leaves to collapse transversely. Young leaves of several broadleaf plants can be uniformly chlorotic, but chlorosis first develops in the basal areas of spinach and tomato.

Excesses of soil-derived Fe are rather rare except in very acidic or flooded soils. When excesses do occur, they usually manifest as a deficiency of Mn. In rice, where toxicity results from flooding and is known as "bronzing," leaves are first covered with small spots that turn brown. Damage from chelated iron sprays, particularly under low light and high nutrient concentrations, is not uncommon.

Symptoms appear as elliptical necrotic spots, which at times can be surrounded by a white chlorotic halo.

Manganese

An Mn deficiency is also commonly manifested as leaf chlorosis. It can be differentiated from Fe shortages by the necrotic spotting or lesions, which appear as black round spots along the midribs and veins of potato leaves; as small brown or orange-tinted necrotic spots close to the major veins and midribs of tomato leaves; as elongated ivory or pale brown spots with blue-green or gray halo areas in oat; as dark brown spots along the veins of barley; as somewhat rectangular spots along midribs and veins of rapidly expanding dwarf French bean leaflets; and as necrotic speckling of chlorotic leaflets in subterranean clover.

Occurrences of excess Mn are much more common than instances of excess Fe, as they occur rather frequently in acidic soils. Alfalfa, apple, the brassicas, cereals, clovers, pineapple, potato, sugar beet, and tomato are especially sensitive to excess Mn. Symptoms of excesses generally appear as a deficiency of Fe, with chlorosis of the leaves being common. Excess Mn tends to produce leaves with irregular mottling between veins. The leaves may have dark brown, purple, or black necrotic spots close to leaf margins. Other symptoms include leaves with white margins (alfalfa), cupped leaves (cabbage and clover), incurling of leaf margins (cauliflower), necrotic spots (cabbage, cauliflower, cereals, citrus, and tung), brittle lower leaves (potato), and distorted laminae (cauliflower).

Molybdenum

Shortages of molybdenum (Mo) are reflected by bright yellow-green interveinal chlorotic mottling of leaves. The chlorosis is followed by the curling, withering, and eventual collapse of leaf margins. Plants grown with nitrate or materials that nitrify quickly tend to show chlorosis first on older leaves. The chlorosis may move on to younger leaves before the plant dies. At times, the symptoms of deficiency appear first on midstem leaves, as is the case with barley, celery, mustard, potato, and tobacco. The deficiency in legumes or nitrogen-fixing nonlegumes is similar to N deficiency. In the brassicas, a deformity that has been characterized as "whiptail" appears mainly on younger leaves. The conditions at first produce leaves with partial blades that may be distorted (strap leaves), but as the plant develops the new leaves may consist only of midribs with necrotic spots.

Although excess Mo can lead to molybdenosis, a serious disease in ruminant animals, Mo toxicity in plants is not observed in the field.

Zinc

Lack of Zn usually leads to small, malformed leaves with irregular mottling and interveinal areas that are yellow-ivory in color. The deficiency is known by many names, all of which are descriptive of the leaf symptoms: "little leaf" for apples; "sickle leaf" for cocoa; "frenching" for citrus; "bronzing" for tung; and "rosette" for subterranean clover, pecan, rubber, and several stone fruits. Distortion of leaves resulting in wavy margins, curling the lamina, and scorching or necrotic spotting, often occurs in pepper, rice, tobacco, and tomato.

As with excesses of other metals, excess Zn is often revealed as an Fe deficiency. In addition, Zn toxicity of barley and several grasses usually leads to rusty brown flecks on the leaves. Leaves so affected die prematurely.

Aluminum

Low pH increases both Mn and Al in soils. Excess Al, which has such a marked effect on roots, has a less noticeable impact on leaves. The normal green color may be modified and appear dull, dark, or olive green. Occasional necrosis may be present. Red purpling of the leaf bases can occur in barley and the collapsing of petioles with the blackening of young leaves in celery.

FLOWERS

Identifying the problems of flowers can help to explain poor yields in many crops. Common problems are: failure to flower or incomplete flowering; poor pollination or fertilization of flowers; and damage to flower parts. An examination of flowers can uncover several problems as outlined below.

Insect damage may be indicated by chewed or aborted blossoms. Webs around flowers, signifying mite damage. The poor set of several crops can result from blossom feeding or attack. Destruction of part of the flower can result in misshapen fruit.

Disease damage is reflected in withered and aborted blossoms resulting in poor set. Spotting may occur which can engulf most of the flower. Streaking of flower color is caused by viral infections. Small fruit may develop even though flowers have been affected by disease but such fruits usually drop prematurely.

Nutritional disorders are responsible for several problems of flowers. Lack of B will reduce flowering and lead to excessive shedding of cotton squares, as well as producing poor fertilization of embryos and poor set of many plants. Premature flower fall may result from low levels of N, B, or Cu. Abnormal anthers and death of the pollen tube before it can penetrate the embryo sac is common in B-deficient plants. Curds of cauliflower can turn brown if B is deficient, or develop a number of interspersed bracts if Mo is limiting. Wilting of flower stems and pedicels of brassicas, some legumes, and tulips shortly before, during, or shortly after the opening of flowers may be the result of low levels of Ca. Premature flower dehiscence may accompany the collapse of pedicels of several plants. Lack of Ca also will cause bract-edge burn on poinsettias and necrotic patches of petunia flowers. Flowers that are pale in color may be lacking N or Mg. Delayed flowering may be due to either a shortage or excess of N.

Spray damage results from the sensitivity of flower parts to osmotic effects. Sprays which may have no visible effect on stems and leaves can injure flowers. Spotting of petals is not uncommon. Spraying Easter lilies with Sunspray oil, a relatively safe insecticide, can cause a split in the corolla with repeated applications and make plants unsaleable.

High temperatures can cause "burning" or a delay in flowering. *Low temperatures* tend to cause the browning of flowers. Temperatures less than 47°F induce poinsettia bracts to turn blue or purplish. Frost can cause petals to turn brown or black and the entire flower along with its leaves to drop.

FRUIT AND SEED

Problems with fruit and/or seed greatly reduce the yield and quality of many crops. Some of the problems are related to the following.

Infectious disease will be indicated by small fruits; fruits that yellow and drop prematurel; distorted fruits; shucks of nuts splitting prematurely; brown spots or patches; lesions; distorted pods with few seeds; gray or brown lesions on pods; pink-to-purple or blemished seeds; and rots in various stages of decomposition.

Viral diseases can result in pitted and deformed fruits, some of which may be mottled.

Insect damage is apparent as distorted fruit; fruits, ears, or pods with holes or tunnels; corky layers or unevenly ripened fruit; and webs around the fruit. Fruit may be punctured by insects laying eggs, which can result in larvae in the fruit. The papaya fruit fly will produce milklike exudates on the fruit as it lays its eggs.

Nutritional disorders are indicated by poor set, premature dropping of fruit, cracking or pitting of fruit, distorted fruit, gum pockets, discolored skin, small or excessively large fruit, poorly colored fruit, withered fruit, fruits that ship or hold up poorly in storage, and poorly developed seeds or distorted seeds.

Individual Nutrients

Nitrogen

Lack of N tends to produce small fruits. Some apple fruits will develop bright red colors which can improve their market value.

Excess N tends to produce large fruit that colors and ships poorly. Excess N is suspected in two serious disorders of tomato ("catfacing" and "graywall"), although there is reason to suspect that lack of B plays a role in catfacing. Graywall appears to require both low K and high N in order to develop.

Phosphorus

Lack of P tends to delay the harvest of fruit, and in the case of pepper, can distort the shape as well.

An excess of P can be manifested as shortages of a few elements: N (small but highly colored fruits), Cu (partially developed or blind ears of cereals), or Zn (aborted fruits or poor seed formation).

Potassium

Plants suffering from a deficiency of K tend to produce small, poorly colored fruits that ship poorly or may shrivel. Rind disorders in orange and mandarin are more prevalent. In excess, K can produce large citrus fruit with coarse rinds. Such fruit will also have less juice, but appreciably more acid. Excess K can induce Ca or Mg deficiency in a number of fruits.

Calcium

Blossom-end rot of tomato, pepper, and watermelon largely results from Ca deficiency. Calcium deficiency produces varying symptoms in different fruits: sunken areas in muskmelon or cantaloupes; cracking in sweet cherries; bitter pit or cork spot in apples; cork spot in pears; "soft-nose" in mangoes; softness in strawberries; premature breakdown in apples; and poor filling in peanuts. Translucence or a water-soaked appearance can be visible in the case of some Ca-deficient applies.

Excesses of Ca can lead to shortages of K or Mg, both of which are associated with lowered sugar content in several fruits.

Boron

Deficiency of B is reflected in a number of fruit disorders: the poor set of many fruits; the pitting, discolored skin, cracking, and corking of apples; the cracked fruit with pockets of gum on the skin or internally of plums and prunes; the uneven fruit development ("hen and chicks") of grapes; the pitting and internal corking of pears; the thickened rind and gum pockets or discolored patches of grapefruits and oranges; the cracking and pale chlorotic skin of strawberries; the hollow heart and dark hollow areas in the center of peanut kernels; the short bent cobs with poor kernel development of corn; and the ring or corky splits around the calyx end and the irregular collapse of tomato fruits.

An excess of B can greatly reduce fruit set in a number of plants. Fruits that do set can be normal in appearance but often ripen prematurely. Scablike protuberances can develop, most of which may be lost as the fruit ripens.

Copper

Deficiencies of Cu resulting in shriveled cereal grains and in tomato may produce symptoms similar to "blossom end rot."

Surpluses of Cu are often revealed as Fe deficiency, which is reflected as reduced set of fruit. Fruit that does set usually fall prematurely. If it does ripen, it may be light in color (citrus).

Iron

Lack of Fe produces off-colors in apple, pear, and tomato fruits. Apples and pears may be chlorotic or have bright red or orange tints. Immature tomatoes will have a silvery green appearance. Instead of being red at maturity, they will often be orange in color.

Manganese

Lack of Mn affects seeds of several plants, causing sunken areas of the cotyledons. This phenomenon in peas, known as "Marsh spot," was one of the first seed disorders identified as an Mn deficiency. Somewhat similar symptoms appear in broad, haricot, and both dwarf and climbing French beans.

Molybdenum

Mild deficiency of Mo can result in cauliflower curds that have many interspersed bracts and poor seed set. Plants developed under conditions of deficient Mo may produce normal looking seeds, but plants developing from such seeds will show early symptoms of deficiency.

Zinc

Fruit and seed production of alfalfa, beans, cereals, citrus, clovers, flax, tomatoes, and walnuts are greatly reduced as a result of Zn deficiency. Fruit size of citrus is reduced and the albedo is rather thick in proportion to pulp. Shells of walnuts fail to harden, tending to remain tough and pliable.

SOIL EVALUATION

It will be necessary to make a careful examination of the soil as well as the plant in order to diagnose most problems. During the period of examination, some idea must be gained as to (1) the adequacy of the water; (2) the efficacy of the drainage; (3) the presence of compaction or hardpans; (4) the sufficiency of soil to support the plant; and (5) the suitability of the soil from the standpoint of temperature. If examined soon after fertilization, it may be possible to locate the placement of fertilizer, which can have a bearing on poor plant development and/or toxicity symptoms. Poor early plant development or failure to obtain a satisfactory stand is often due to placement of fertilizer too close to the seed or directly under the emerging plant. Plant damage as exhibited by "burnt" roots and leaves with "burnt" edges may at times also be due to sidedressing fertilizer placed too close to the roots.

A preliminary examination of the soil profile can be made with a probe or auger, but often a more detailed examination will be necessary. In such cases, the entire profile needs to be exposed (see Chapter 1). On examination, observations should be made regarding depth and conditions of the roots, depth of the different horizons, extent of compaction at various levels, and the presence of stratification (abrupt changes in texture). Samples collected every 6-8 inches and analyzed for texture, pH, conductivity, Al, and nutrients can help identify the causes of many problems.

Examination of soil or soilless mixtures in pots is easily made in most situations by carefully inverting the pot and tapping the base lightly to remove the entire plant.

Soil Texture

The soil's texture needs to be known since many of the properties of the soil are influenced by it. An estimate of texture can be made in the field by feel and this can help in early diagnosis of a problem. Determination in a laboratory of profile samples collected during the examination often can be beneficial in determining whether stratification is a factor in water movement.

Compaction

Compaction, which often limits crop production, can be assessed by penetration with a Hofer tube or more accurately by penetrometers. (See Measuring Compaction in Chapter 2.)

The examination of roots exposed in a profile or carefully removed by shovel will help corroborate findings revealed by a Hofer tube or penetrometers. Roots in a normal soil will penetrate deeply with little or no kinks; roots in compacted soils tend to be short, kinky, and show abrupt changes from a vertical to a horizontal orientation.

Water Sufficiency

A quick appraisal of current water sufficiency can be made by testing the feel of soil samples collected at various depths (Table 1.1).

Exposure of the root system obtained by carefully lifting the plant or by digging a trench alongside the plants to reveal the presence of the majority of roots will help diagnose whether most roots are getting enough water. Soil samples of the different layers evaluated by feel can quickly provide information on the current status of water.

Sufficiency of Drainage

Examination of soil removed by deep probes or exposed by digging trenches can reveal excess water harmful to many plants. Such excess water coming sometime after a rain or irrigation is highly indicative of problems. The water content can be estimated by feel or determined by soil analysis. The determination along with root position and appearance can signify whether excess water is a problem.

The color of subsoils is also indicative of drainage. Soils with good drainage tend to have a uniform color or a gradual change in color with increasing depth. The normal red and brown colors of highly weathered subsoils tend to become paler if moisture is high and in extreme cases of excess water the soil will be gray or blue. Under conditions of extended wet periods followed by dry periods, such subsoils will tend to have a mottled appearance.

WATER

Adequacy

The shortage or excess of water as indicated by soil sampling can be confirmed to a certain extent by crop appearance. Plants failing to receive enough water show limited growth, have flaccid gray leaves, and tend to wilt for long periods during the day.

Unfortunately, plants will also show extreme wilt under conditions of excess water. Excess water limits oxygen around roots, making it difficult for roots to absorb the water. The existence of such a situation is easily determined by examination of the soil as well as the plant.

Elevated plant temperatures, which can be measured quickly by a portable hand infrared thermometer, are indicative of a plant's water status. The rise in temperature results from stomata closing. Unfortunately for diagnostics, stomata may close because of several stress factors, such as disease, root pruning, or fertility problems. Infrared temperature readings need to be coupled with weather data and crop observations to be more meaningful.

Tensiometer data, particularly if it is graphed, can be of great assistance in determining whether plants have been receiving too little or too much water. Low tension readings (less than 10 centibars for sands; less than 30 centibars for heavy soils) for long periods can signify shortages of soil oxygen and, unless on a continuous fertigation, shortages of the leachable elements (N, K, and

Mg) as well. Low tension readings (less than 40 for sands; less than 60 for heavy soil) close to harvest can indicate problems with shipping quality of a number of fruits. On the other hand, high tensions, coming during early growth and particularly as fruits are set, are indicative of insufficient water.

Determining moisture by one of the newer moisture probes can provide immediate information as to moisture status. A combination salinity/moisture probe manufactured by Aquaterr Instruments, Freemont, California is combined with a hand-held data logger and a computer program to monitor field conditions.

Chapter 18

Appraising the Problem: Diagnoses

The cause of a plant problem often will be apparent from the various observations described in the preceding chapter. Weed competition damage, for example, can be quickly evaluated by noting the numbers of weeds and their size. Adequate control depends on properly identifying the weed, which can be done by comparing illustrations appearing in many different publications.

Problems may arise in using symptoms alone to diagnose pathogenic diseases, nutritional disorders, or poor growth caused by environmental factors because of similar symptoms caused by different agents. At times, the problem can be resolved by comparing the symptoms caused by different agents simultaneously. In many cases, observations can narrow the possible causes, and a tentative identification can be made by carefully comparing symptoms caused by each agent. Several publications that discuss the damage caused by many different agents can be found among the Additional Readings at the end of this section.

Outlines for on-farm identification of the problem's cause are presented in this chapter under several potential agents. But if the problem cannot be identified at the farm, suitable samples must be sent to public or private agencies for identification.

Several states maintain diagnostic clinics where plant samples may be sent for identification. A few that the author is familiar with are the Plant and Pest Diagnostic Clinic at Ohio State University and several Florida Extension Plant Disease Clinics (FEPDC) located in Gainesville, the Tropical Research and Education Center in Homestead, and the North Florida Research and Education Center (NFREC-PDC) in Quincy. A number of commercial laboratories are also capable of diagnosing insect, infectious disease, nematode, and nutritional problems.

Infectious diseases can at times be diagnosed by the symptoms. Descriptions of symptoms of many different diseases can be found in many publications. Some of those that the author has found useful are presented under Additional Readings.

Since many of these symptoms can be caused by agents other than infectious diseases, it is helpful if the presence of disease can be corroborated by other means. A hand lens of 10-20 magnification often can help rule out some of the possibilities and allow for preliminary identification if a disease is present. For example, the presence of mycelial growth and/or spores can be determined with a hand lens and is indicative of a fungal infection. The color of the mycelia and spores may help identify the particular organism. Both Rhizopus and Botrytis infections found in poinsettia propagation have grayish mycelia, but the Rhizopus spores are black, while those of Botrytis are gray.

Careful examination of all parts of the plant are necessary for a preliminary diagnosis. Roots and stems need to be cut open and examined, and the upper and lower sides of leaves, flowers and fruits should also be inspected. While such diseases as powdery mildew, downy mildew, rusts, leaf spots, bacterial blight (angular leafspot), Helminthspoium brown leafspot, and Alternaria blight can be preliminarily identified by leaf symptoms, a number of diseases, such as Fusarium wilt, Pythium stem rot, Verticillum wilt, bacterial wilt, Rhizoctonia root canker and Phytophthora root rot, are revealed by stem and root symptoms. A few of the latter diseases can produce leaf symptoms, but discoloration of xylem (visible when root and stems are split open) or necrosis of the stem at the soil line are often the keys to identification.

Knowing the disease symptoms of crops commonly grown in an area can aid in a preliminary identification.

Positive identification usually requires much higher magnification than is possible with a hand lens. A compound microscope of 100-1000 power is often needed. This type of identification also requires highly trained personnel who may use isolation and special culturing techniques in addition to microscopic examination to make the diagnosis. For many situations, personnel engaged in routine IPM examinations may be able to make the diagnosis, or it may be accomplished by one of the newer identification kits. (See Chapter 10.)

If identification cannot be made at the farm, it will be necessary to deliver or send a suitable sample to either the state extension service or a commercial laboratory offering identification services. To obtain reliable identification, the sample must be collected and handled in special ways in order for the sample to be serviceable. Directions for collecting and submitting samples for disease evaluation are outlined in Chapter 10.

Positive identification of viral caused damage also requires considerable skill and laboratory facilities. Entire small plants or affected parts of older plants need to be submitted.

Insect damage is relatively easy to confirm if the insect is found attacking the plant. Undersides of the leaf are the common sites for many insects. Placing a sheet of white paper under leaves and striking them sharply will cause thrips or mites to fall on the paper, making it easier to locate them. A drop cloth with handles placed between rows of plants is more suitable for collecting and counting larger insects shaken from the plant. Flying insects can be collected by using sweep nets or with various traps. (See Chapter 10.)

Insects may be present on any part of the plant. Roots are often overlooked, but the larvae of a number of insects may be causing damage and the roots should be examined carefully.

The presence of an insect does not always mean that it is causing the damage. Some insects may simply be resting on the plant. The type of damage must be positively associated with the insect in question and this requires identification. Descriptions are available in many different publications, some of them are available for free or at a nominal cost from state extension services. Some of the more helpful publications the author has used are listed under Additional Reading at the end of this section.

Identification can be made by noting the insect's feeding habits as well as its appearance and comparing these with various photographs, line drawings, and descriptions offered in several publications listed under Additional Reading at the end of this chapter. Using a hand lens of 10-20 power greatly aids in identifying the smaller pests (thrips, mites, and scales) and the eggs, nymphs, or young larvae of many pests.

The type of damage and its location are helpful in making diagnoses. For example, a stippled mottling pattern on foliage is typical of spider mite damage. Spider mites may cause palm leaves to have a bleached appearance. Trained personnel conducting IPM usually can make the identification. If services are not available, unidentified insects can often be sent to state extension services for identification. Contact the local county agent for directions for mailing, or follow the directions provided in Chapter 10. Some commercial laboratories also can help make the identification.

Nematode damage from root-knot organisms are relatively easy to diagnose in the field. Damage from other types may be confused with symptoms caused by disease, nutritional problems, or insufficient oxygen due to compaction or flooding. Positive identification will depend upon evaluating a sample that can often be submitted to state extension services or commercial laboratories. Directions for routine sampling in Chapter 10 can be the basis for collecting a sample for damage assessment. The major restriction in sampling for nematode diagnosis and identification is that sampling needs to be limited to areas among the roots of damaged plants. Collecting the sample around plants that show symptoms but are not yet dead is desirable.

Herbicide damage can occur either from overdoses or from carryover from previous applications. Carryover is a greater problem if a plant that is sensitive to the herbicide follows the treated plants, especially if soil moisture has been low. Damage attributed to herbicides based upon leaf symptoms or even the death of plants can often be confirmed by collecting a soil sample from the damaged area and evaluating it by bioassay, immunoassay or chromatography. (See Chapter 11.)

Nutritional problems can be diagnosed by plant symptoms, but because symptoms linked to nutrient imbalances could have other causes, the diagnosis needs to be confirmed by other means. Confirmation can be obtained through soil and leaf analyses. Some of the problems can be addressed or their effects can at least be eliminated by tests performed at the farm. Suitability of pH can be diagnosed by a portable soil pH meter. The presence of excess salts can easily be determined by a portable conductivity meter. Low conductivity readings can also signify general low macronutrient levels.

Tissue tests run on the farm can help diagnose problems of the major elements, particularly for N and K. (See Sap Analysis and Table 5.3 in Chapter 5.) The results from plant rapid tests can function as a basis for starting foliar feeding, but it is usually desirable to obtain the complete results of an analysis carefully selected soil and plant samples before making soil fertilizer applications.

The diagnosis of nutrient excesses or shortages often cannot be made without more complete analyses of both soil and leaf, which will have to be done at a laboratory. Directions for collecting and shipping soil samples were given in Section I and those for plants in Section II. Commercial laboratories or those operated by state extension services can provide the necessary analyses.

Collection of soil and leaf samples according to standard sampling procedures may not reveal nutritional problems. The process can be greatly enhanced if two separate soil samples are collected, one from soil supporting normal plants and the other from soil where plants are doing poorly, along with two leaf samples collected from normal and problem plants. Often, the symptoms are present on leaves or other plant parts that are not the basis for standard tests. Therefore, in selecting tissue, it is very important to sample both normal and abnormal tissue of the same age and position on the plant.

Environmental problems are often apparent from visual symptoms, but it is helpful if they can be confirmed by more positive means.

The suitability of temperature, humidity, and CO_2 levels in greenhouse operations can be evaluated from recorded data. If these are not available, current values can be obtained from immediate measurements. If possible, set up recording devices to evaluate several parameters. While such records cannot be helpful in distinguishing single past causes of poor growth, they can be useful if current conditions are still not ideal.

Excess levels of gases that are toxic to plants in protected structures or transport vehicles can often be assessed by color detecting tubes (Chapter 15) or by submitting air samples for gas chromatography.

Confirmation of damage to plants due to an excess of ethylene in protected structures or transport vehicles also can be obtained by using color-detecting tubes or chromatography of air samples. A bioassay using young vigorous tomato or African marigold plants with three to four fully developed leaves placed in suspected areas may also be a practical means of assessing ethylene excess, since leaves will exhibit epinasty (downward bending) within 24 to 48 hours of exposure. The practicality of this bioassay is dependent on the availability of young potted plants.

Assessing damage due to atmospheric conditions in the field is much more difficult. Accurate thermometers and temperature recorders can be helpful during excessively cold or hot periods. At times, data from local weather stations can also be useful in evaluating past environmental conditions. But often, an appraisal is made from indirect evidence of symptoms.

The presence of toxic levels of air pollutants in the field is usually based not only on plant symptoms, but also on the proximity of the source of pollution and whether affected plants are located downwind from the offending source. Sources of several pollutants and the normal radius in which plants may be affected are outlined below.

Sulfur dioxide. Sources of SO_2 are power stations using fossil fuels and industrial plants involved in ore smelting, steel manufacturing, and petroleum refining. Local problems can arise from large space heaters using fossil fuels, refuse burning, and burning of coal refuse piles. Volcanoes and fumaroles also can emit toxic levels of SO_2.

Most of the SO_2 damage to plants is confined to an immediate area close to the source, with damage decreasing as distance from the source increases. But if the industrial source is emitted in tall smokestacks, damage may occur 5 to 10 miles from the source. Sulfur dioxide emitted from volcanoes may also do damage at considerable distances from the source.

Hydrogen fluoride or silicon tetrafluoride (SiF_4). Both materials are emitted when fluoride-containing rocks and minerals are heated or chemically treated (as is the case in the manufacture of aluminum, phosphatic fertilizers, glass, brick, steel, and chemicals) or when coal is burned to produce heat or steam. Hydrogen fluoride is also produced by fumaroles and volcanoes.

Most damage occurs within a few miles of the source. In some cases, if the gases are emitted by tall smokestacks or if the industrial plume is not dispersed, as happens under certain mountain and valley conditions, damage to plants can occur 10 to 20 miles from the source.

Ozone. Ozone accumulates to phytotoxic levels as sunlight reacts with nitrogen oxides in the presence of hydrocarbons. Urban areas are the principle sites for production of O_3, although small amounts are produced by lightning. The latter source is of limited concern, except with plants produced at high altitudes. Most damage will be confined to areas close to urban development or roadsides with heavy traffic.

Peroxacetyl nitrate. The common silvering, glazing, or bronzing of leaves resulting from the phytotoxic effects of PAN are also primarily confined to urban areas. Damage that is primarily confined to the lower sides of actively growing leaves provides evidence of phytotoxicity caused by PAN. Damage is usually worse on vigorous plants grown with ideal moisture and fertility levels.

Confirmation of PAN injury on the basis of leaf symptoms is made difficult because of variable symptoms or no visible symptoms. Several plants exposed to low levels of PAN for short periods will exhibit chlorosis on the leaves, rather than the typical glazing, silvering, or bronzing. Others may show no visible signs of injury, even though their yields will be adversely affected.

Additional Reading

The reader is referred to the citations under Additional Reading at the end of Section IV, many of which give excellent descriptions of pests and the damage they cause. The following, some of which also describe pest damage, can be helpful in evaluating nonpest problems as well.

NUTRITIONAL DISORDERS

Bennett, W. F. (ed.). 1983. *Nutrient Deficiencies and Toxicities in Crop Plants*. APS Press, St. Paul, MN 55121-2097. Good descriptions and photos of nutrient deficiencies and excesses in a number of plants.

Bergmann, W. 1992. *Nutritional Disorders of Plants–Development, Visual and Analytical Diagnosis*. (Translated from the German by B. Patchett.) This is a large book (741 pages) and expensive ($214), but has 945 color pictures of many different plants exhibiting symptoms of deficiencies and excesses of macro and micronutrients. In addition, a number of illustrations depict symptoms displayed by plants exposed to gases, acid rain, herbicides, and excess salts.

Bould, C., E. J. Hewitt, and P. Needham. 1984. *Diagnosis of Mineral Disorders in Plants*. Her Majesty's Stationery Office, London. Excellent descriptions and illustrations of nutrient deficiencies and excesses of many plants.

Chapman, H. D. (ed.). 1966. *Diagnostic Criteria for Plants and Soil*. Division of Agricultural Sciences, University of California. Contains descriptions and photographs of nutritional disorders plus interpretative data of plant analysis.

Weir, R. G. and G. C. Cresswell. 1993. *Plant Nutrient Disorders I–Temperate and Subtropical Fruit and Nut Crops*. Inkata Press,

North Ryde 2113, Australia. Contains a number of color plates of nutrient deficiency and toxicity symptoms in fruit trees, as well as those of injuries resulting from exposure to air pollution, adverse sun and wind conditions, and various sprays.

WEEDS

Agri-Growth Guides to Herbicide Injury. Agri-Growth Research Inc., Hollandale, MN 56045-9799. Injury symptom guides for corn, cotton, small grains, and soybeans are available.

Muenscher, W. E. 1981. *Weeds.* Cornell University Press, Ithaca, NY.

Rice, R. P. 1986. *Nursery and Landscape Weed Control Guide.* Thompson Publications, P.O. Box 9335, Fresno, CA 93791.

GENERAL DIAGNOSIS

Chase, A. R. and R. T. Poole. 1986. Troubleshooting guide to foliage. *Greenhouse Grower* 4:(6) p. 54-55, (7) p. 104-105, (10) p. 68-69, (11) p. 24, 25, and 27.

Flint, M. L. 1990. *Pests of the Garden and Small Farm.* University of California Agricultural and Natural Resources, Oakland, CA 94608. This book, which emphasizes alternate methods of pest control, has 250 color photographs, 28 black-and-white photographs, and 90 line drawings, many of which can help in diagnosis.

Henley, R. W. 1981. Diagnosing Plant Disorders. In *Foliage Plant Production.* N. J. Joiner (ed.). Prentice-Hall, Inc., Englewood Cliffs, NJ 07632.

Marlatt, R. B. 1980. *Noncontagious Diseases of Tropical Foliage Plants.* Bull 812 IFAS, University of Florida, Gainesville, FL. Presents descriptions and photographs of nutritional deficiencies and spray damage as well as descriptions of stress damage from growth regulators, humidity, temperature, wood preservatives, light, and water.

Skelly, J. M., D. D. Davis, W. Merrill, E. A. Cameron, H. D. Brown, H. D. Drummond, and L. S. Dochinger. 1987. *Diagnosing Injury to Eastern Forest Trees.* Penn State College of Agriculture, University Park, PA. Contains color photographs which are helpful in diagnosing the problems of forest trees.

Smith, M. D. (ed.). 1983. *Ortho Problem Solver.* Ortho Information Services, Chevron Chemical Co., 575 Market St., San Francisco, CA 94105. An excellent treatise featuring problems of gardens and home landscapes.

Yepsen, R. B. (ed.). 1984. *Encyclopedia of Natural Insect and Disease Control.* Rodale Press, Emmaus, PA. A rather complete book of disease and insect activity affecting a great number of crops. Includes descriptions, line drawings, and color photographs. Defines numerous terms related to insects, diseases, and controls, with an emphasis on nonchemical or alternate control methods.

SECTION VII:
APPENDIXES

Appendix 1

Abbreviations and Symbols Used in This Book

a = acre
a.i. = active ingredient
a.i./a = active ingredient per acre
Al = aluminum
Al^{3+} = aluminum ion
B = boron
C = carbon
(°C) = degree centigrade
Ca = calcium
Ca^{2+} = calcium ion
CCE = calcium carbonate equivalent
cc = cubic centimeters
CEC = cation exchange capacity
Cl = chlorine or chloride
Cl^- = chloride ion
C/N = carbon/nitrogen ratio
Cu = copper
Cu^{2+} = copper ion
EC = electrical conductivity
(°F) = degree Fahrenheit
F = fluorine
F^+ = fluoride ion
Ft = foot or feet
g = gram
gr = grain
gal = gallon
H = hydrogen
H^+ = hydrogen ion
$H_2BO_3^-$ = borate ion

H_2CO_3 = carbonic acid
HCO_3^- = bicarbonate ion
H_2O = water
HSO_4^- = sulfate ion
in or ʺ = inch or inches
K = potassium
K_2O = potassium oxide or potash
lb = pound
lb/a = pound or pounds per acre
meq/100 g = milliequivalents per 100 grams
Mg = magnesium
Mg^{2+} = manganese ion
ml = milliliter or cubic centimeter
Mn = manganese
Mn^{2+} = magnesium ion
mmhos = millimhos
Mo = molybdenum
MoO_4^{2-} = molybdate ion
mph = miles per hour
N = nitrogen
Na = sodium
Na^+ = sodium ion
NH_3 = ammonia
NH_4^+ = ammonium ion
$NH_4–N$ = ammonium nitrogen
N_2O = nitrous oxide
NO_2 = nitrogen dioxide
$NO_2–N$ = nitrite nitrogen
NO_3^- = nitrate ion
$NO_3–N$ = nitrate nitrogen
O = oxygen
OH^- = hydroxyl ion
OM = organic matter
oz = ounce
P = phosphorus
P_2O_5 = phosphorus pentoxide (phosphate)
ppb = parts per billion
ppm = parts per million
psi = pounds per square inch
pt = pint
qt = quart

S = sulfur
SAR = sodium absorption ratio
Si = silicon
SO_2 = sulfur dioxide
$SO_4{}^{2-}$ = sulfate ion
sq ft = square feet
ton/a = ton or tons per acre
Zn = zinc
Zn^{2+} = zinc ion

< = less than
> = greater than

Appendix 2

Important Soil, Plant, Water, and Fertilizer Elements and Ions with Symbols, Atomic, and Equivalent Weights

Element	Symbol	Atomic Wt.[1]	Ion	Symbol and Valence	Atomic Wt.[1]	Equivalent Wt.[1]
Aluminum	Al	27	Aluminum	Al^{3+}	27	9
Boron	B	11	Borate	$H_2BO_3^-$	61	61
Calcium	Ca	40	Calcium	Ca^{2+}	40	20
Carbon	C	12	Carbonate	CO_3^{2-}	60	30
			Bicarbonate	HCO_3^-	61	61
Chlorine	Cl	35.5	Chloride	Cl^-	35.5	35.5
Copper[2]	Cu	64	Cuprous	Cu^+	64	64
			Cupric	Cu^{2+}	64	64
Fluorine	F	19	Fluoride	F^+	19	19
Hydrogen	H	1	Hydrogen	H^+	1	1
			Hydroxyl	OH^-	17	17
Iron[3]	Fe	56	Ferrous	Fe^{2+}	56	28
			Ferric	Fe^{3+}	56	19
Magnesium	Mg	24	Magnesium	Mg^{2+}	24	12
Manganese[4]	Mn	55	Manganous	Mn^{2+}	55	27.5
			Manganic	Mn^{3+}	55	18
Molybdenum	Mo	96	Molybdate	MoO_4^{2-}	160	80
Nitrogen	N	14	Ammonium	NH_4^+	18	18
			Nitrate	NO_3^-	62	62
			Nitrite	NO_2^-	46	46
			Urea	NH_2^-	16	16
Oxygen	O	16	Oxygen	O^{2-}	16	16
Phosphorus	P	31	Phosphate	$H_2PO_4^-$	97	97
			Phosphate	HPO_4^{2-}	96	48
Potassium	K	39	Potassium	K^+	39	39
Silicon	Si	28	Silicate	$HSiO_3^-$	77	77

Sodium	Na	23	Sodium	Na^+	23	23
Sulfur	S	32	Sulfate	SO_4^{2-}	96	48
Zinc	Zn	65	Zinc	Zn^{2+}	65	32.5

[1] To closest whole or half number.

[2] It is believed that the cupric ion is important for plant nutrition.

[3] Both ferrous and ferric ions are important in soil and plant chemistry, but the ferrous form is required by plants.

[4] Manganese can exist in five different valence forms, but 2+ and 3+ are the primary manganese ions in soil chemistry processes, with the 2+ form being the important one for oxidation/reduction and photochemical processes in plants.

Appendix 3

Common and Botanical Names of Plants

Common Name	Botanical Name
Aglaonema or Chinese evergreen	*Aglaonema commutatum* Schott
Alfalfa or Lucerne	*Medicago sativa* L.
Allamanda	*Allamanda cathartica* L.
Almond	*Prunus amygdalus* Batsch
Alstroemeria	Alstroemeria sp.
Anthurium	*Anthurium andraeanum* Linden
Aphelandra or Zebra plant	*Aphelandra squarrosa* Nees
Apple	*Malus* sp.
Apricot	*Prunus armeniaca* L.
Aralia, false	*Aralia dizygotheca*
Arborvitae	*Thuja orientalis* L.
Ash, European	*Sorbus aucuparia* L.
Ash, White	*Fraxinus americana* L.
Asparagus	*Asparagus officinalis* L.
Asparagus fern	*Asparagus retrofractus* L.
Asparagus, Myers or Sprengeri	*Asparagus densiflorus* (Kunth)
Aspen, Trembling	*Populus tremuloides aurea* Michx.
Avocado	*Persea american* P. Mill
Azalea	*Rhododendron indicum* L.
Azalea, Indian hybrid	Rhododendron sp. hybrids
Baby's breath	*Gypsophila paniculata* L.
Banana	Musa sp.
Barley	*Hordeum vulgare* L.
Basswood or Linden	*Tilia* sp.
Bean, Broad, Fava, or Horse	*Vicia faba* L.
Bean, Snap	*Phaseolus vulgaris* L.

Beech	*Fagus grandifoila Ehrh.*
Beet, Sugar	*Beta saccharifera*
Beet, Table	*Beta vulgaris* L.
Begonia, Rieger	*Begonia x hiemalis* Fotsch
Begonia, Wax leaf	*Begonia x semperflorens-cultorum*
Birch, Black	*Betula nigra* L.
Birch, White	*Betula papyrifera* Marsh
Birch, Wire	*Betula populifolia* Molsn
Birch, Yellow	*Betula alleghaniensis, B. lutea,* or *B. verrucosa*
Bird of paradise	*Strelitzia reginae Ait.*
Birds-foot trefoil	*Lotus corniculatus* L.
Blueberry, Highbush	*Vaccinium corymbosum* L.
Blueberry, Rabbit eye	*Vaccinium ashei* Reade
Bougainvillea	*Bougainvillea* sp.
Boxwood, Japanese	*Bucus macrophylla var. japonica*
Broccoli	*Brassica oleacea, Botrytis* group
Bromeliad	*Aechmea fasciata Ruiz & Pajon*
Brussel sprouts	*Brassica oleracea, Gemmifera* group
Cabbage	*Brassica oleracea, Capitata* group
Cabbage, Chinese	*Brassica rapa L. Chinensis* group
Cactus, Christmas	*Schlumbergera bridgesii* Lem.
Caladium	*Caladium sp. Vent*
Calathea	*Calathea sp. G.F. W., Mey*
Cantaloupe or Muskmelon	*Cucumis melo L. Reticulatus group*
Carissa or Natal plum	*Carissa grandiflora E. H. Mey*
Carnation	*Dianthus caryophyllus* L.
Carrot	*Daucus carota* L.
Cashew	*Anacarium occidentaled* L.
Cassava	*Manihot esculenta crantz*
Cauliflower	*Brassica oleracea, Botrytis* group
Cedar, Western red	*Thuja plicata Donn ex D. Don*
Celery	*Apium graveolens var. dulce* (Mill.) Pers.
Chalkas or Orange jasmine	*Murraya paniculata* (L.) Jack
Cherry, Black	*Prunus serotina Ehrh.*
Cherry, Choke	*Prunus virginiana* L.
Cherry, Pin	*Prunus pennsylvanica* L.
Cherry, Sour	*Prunus cerasus* L.
Cherry, Sweet	*Prunus avium* L.

Chrysanthemum	*Chrysanthemum x morifolium* Ramat
Clover, Alsike	*Trifolium hybridum* L.
Clover, Ladino or White	*Trifolium repens* L.
Clover, Red	*Trifolium pratense* L.
Clover, Subterranean	*Trifolium subterraneum* L.
Coco-plum	*Chrysobalanus icaco* L.
Cocoa	*Theobroma cacao* L.
Coffee	*Coffea arabica* L.
Collards	*Brassica oleracea, Acephala group*
Corn or Maize	*Zea mays* L.
Corn, Sweet	*Zea mays var. rugosa Bonaf.*
Cotoneaster	*Cotoneaster apiculatus* Rehd. & E. H. Wils
Cotton	*Gossypium hirsutum* L.
Cranberry	*Vaccinium macrocarpon Ait*
Croton	*Codiaeum variegatum* (L.) Blume
Crown-of-thorns	*Euphorbia milii Des Moulons*
Crownvetch	*Coronilla varia* L.
Cucumber	*Cucumis sativus* L.
Currant, Black	*Ribes nigrum* L.
Cyclamen	*Cyclamen persicum* Mill.
Desmodium, Greenleaf	*Desmodium intortum* (P. Mill) Urban
Dieffenbachia	*Dieffenbachia maculata* Lodd.
Dogwood	*Cornus alba* L. and *C. racemosa* Lam.
Dracaena godseffiana	*Dracaena surculosa* Lindi.
Dracaena, Janet Craig	*Dracaena deremensis 'Janet Craig'*
Dracaena, Corn plant	*Dracaena fragrans 'Massangeana'*
Dracaena, Sanders	*Dracaena Sandrana*
Dracaena, Reflexa	*Dracaena thalioides*
Dracaena, Warnecki	*Dracaena deremensis 'Warneckii'*
Eggplant	*Solanum melongena* L.
Endive or Escarole	*Cichorium endivia* L.
Eugenia	*Eugenia sp.*
Eucalyptus	*Eucalyptus deglupta* or *E. grandis*
Eucalyptus, Southern blue-gum	*Eucalyptus globulus* Labill.
Euonymus	*Euonymus alatas* (Thumb.) Siebold
Fava beans	*Vicia faba* L.
Fern, Birds-nest	*Asplenium nidus* L.

Fern, Boston	*Nephrolepis exaltata* 'Bostoniensis'
Fern, Leatherleaf	*Rumohra adiantiformis* (Forst f.) Ching
Fern, Maidenhair	*Adiantum pedatum* L.
Fern, Pteris	*Pteris sp.* L.
Fescue, Tall	*Festuca arundinacea* (Schreb.) Wimm
Ficus, Benjamin	*Ficus benjamina* L.
Ficus, Nitida	*Ficus nitida*
Ficus, Decora or Rubber tree	*Ficus elastica* 'Decora'
Ficus, Fiddleleaf	*Ficus lyrata* Warb.
Fig	*Ficus carica* L.
Fir, Amabilis	*Abies amabilis*, Dougl. ex Forbes
Fir, Balsam or Alpine	*Abies balsamea* (L.) P. Mill.
Fir, Douglas	*Pseudotsuga menziesii* (Mirbel) Franco
Forsythia	*Forsythia x intermdia* Zab.
Gardenia	*Gardenia jasminoides* Ellis
Garlic	*Allium sativum* L.
Geranium	*Pelargonium x hortorum Bailey*
Gerbera or Transvaal daisy	*Gerbera jamesonii*
Gladiolus	*Gladiolus x hortulanus* L. H. Bailey
Gloxinia	*Sinningia speciosa* Clodd.
Grape	*Vitis vinifera L., V. labbruska* L., and hybrids
Grape, Muscadine	*Vitis rotundifolia* Michx.
Grapefruit	*Citrus x paradisi* McFaddy
Grass, Bahia	*Papsalum notatum* Flügge
Grass, Bermuda	*Cynodon dactylon* (L.) Pers.
Grass, Bluejoint	*Calamagrostis canadensis* Beauv.
Grass, Brome	*Bromus inermis* Leyss.
Grass, Coastal Bermuda	*Cynodon dactylon* (L.) Pers.
Grass, Creeping bent	*Agrostis palustris*
Grass, Crested wheat	*Agropyron desertorum* Schultes
Grass, Kentucky blue	*Poa pratensis* L.
Grass, Orchard or Cocksfoot	*Dactylis glomerata* L.
Grass, Pangola	*Digitaria decumbens* Stent.
Grass, Perennial rye	*Lolium perenne* L.
Grass, Prairie	*Bromus unioloides* HBK
Grass, Sorghum-sudan	*Sorghum sudanese* Piper

Grass, Sudan	*Sorghum vulgare*
Grass, St. Augustine	*Stenotaphrum secundatum* (Walte.) Kuntz
Grass, Switch	*Panicum virgatum* L.
Grass, Tall fescue	*Festuca arundinacea* Schreb
Grass, Zoysia	*Zoysia matrella* (L.) Merr.
Hawthorn	*Crataegus phaenopyrum* (L.F.) Medic.
Hazelnut	*Corylus avellana* L.
Hemlock Western	*Tsuga heterophylla* (Raf.) Sarg.
Hibiscus	*Hibiscus rosa-sinensis* L.
Holly, American	*Ilex opaca Ait.*
Holly, Chinese	*Ilex cornuta* Lindl. & Paxt.
Holly, Japanese	*Ilex crenata* Thumb.
Honeylocust, Moraine	*Gleditsia triacanthos var. inermis*
Horse chestnut	*Aesculus hippocastanum* L.
Horseradish	*Armoracia rusticana* P. Gaertn., B. Mey, and Scherb.
Hydrangea	*Hydrangea macrophylla* Thumb.
Ixora	*Ixora coccinea* L.
Ivy, English	*Hedera helix* L.
Jasmine, Wax	*Jasminum simplicifolium* G. Forst
Juniper	*Juniperus* sp.
Juniper, Bar Harbour	*Juniperus horizontalis* Moench
Juniper, Pfitzer compacta	*Juniperus chinesis, Pfitzerana compacta*
Kalanchoe	*Kalanchoe Blossfeldiana* Poelln.
Kale	*Brassica oleracea, L. Acephala* group
Kenaf	*Hibiscus cannabinus* L.
Kohlrabi group	*Brassica oleracea, L. Gongylodes*
Larch, Japanese	*Larix Kaempferi* (Lamb.)
Lemon	*Citrus limon* L.
Leea or Hawaiian holly	*Leea coccinea* Planch.
Lettuce	*Lactuca sativa* L.
Ligustrum	*Ligustrum* sp. L.
Lily, Day	*Hemerocallis* sp.

Lily, Easter	*Lilium longiflorum* Thumb.
Lime, Persian	*Citrus aurantiifolia* 'tahiti' (Christa)
Lipstick plant	*Aeschynanthus pulcher* (Blume)
Liriope	*Liriope Muscari* L. H. Bailey
Macadamia	*Macadamia ternifolia* F. J. Muell
Mandarin or tangerine	*Citrus reticulta Blanco*
Malpighia	*Malpighia* sp.
Mandevilla, Dipladenia	*Mandevilla splendens* (Hook)
Maple, Black	*Acer nigrum*
Maple, Mountain	*Acer spicatum* Lam.
Maple, Red	*Acer rubrum* L.
Maple, Striped	*Acer pennsylvanicum* L.
Maple, Sugar	*Acer saccharum* L.
Mango	*Mangifera indica* L.
Maranta or Rabbit's foot	*Maranta Kerchoviana*
Marble queen or Pothos	*Scindapsus aureus*
Millet	*Setaria italica* (L.) Beauv.
Monkey puzzle tree	*Araucaria bidwillii*
Nephytis	*Syngonium podophyllum Schott*
Norfolk island pine	*Araucaria excelsia or heterophylla*
Oak, California live	*Quercus agrifolia* Nee
Oak, Pin	*Quercus palustris Muench.*
Oak, Red	*Quercus rubra* L.
Oak, Valonea	*Quercus ithaburensis*
Oat	*Avena sativa* L.
Olive	*Olea europaea* L.
Olive, Black	*Bucida bucerus* L.
Onion	*Allium cepa* L. Cepa group
Orange, Navel and Valencia	*Citrus sinensis* L.
Orchid, Cattleya	*Cattleya* sp.
Orchid, Cymbidium	*Cymbidium* sp.
Orchid, Ladyslipper	*Cypripedium* sp.
Orchid, Phalenopsis or Moth	*Phalaenopsis* sp.
Palm, Areca	*Chrysalidocarpus lutescens* H. Wendl.
Palm, Bamboo or Seifritzii	*Chamaedorea erumpens* H. E. Moore
Palm, Chamadorea or Parlor	*Chamaedorea elegans* Mart.

Palm, Kentia	*Howea Forsterana* C. Moore & F. J. Muell.
Palm, Oil	*Elaeis guineensis* Jacq.
Palm, Ponytail	*Beaucarnea recurvata* Lem.
Palm, Rhapis or Lady	*Rhapis excelsa* Thumb.
Palm, Roebelini	*Phoenix roebelenii* O'Brien
Papaya	*Carica papaya* L.
Pea, English	*Pisum sativum* L.
Pea, Southern or Black-eyed	*Vigna unguiculata* L.
Peach	*Prunus persica* L.
Peanut	*Arachis hypogaea* L.
Pear	*Pyrus communis* L.
Pecan	*Carya illinoinesis* L.
Peperomia	*Peperomia obtusifolia* L.
Pepper	*Capsicum annum var. annum* L.
Philodendron cordatum or	*Philodendron scandens oxycardium* Heart-leaf
Philodendron hastatum	*Philodendron hastatum* C. Koch & H. Sello
Philodendron panduriforme	*Philodendron panduriforme* HBK
Philodendron pertussum or Monstera	*Monstera deliciosa. Liebm.*
Philodendron selloum	*Philodendron Selloum* C. Koch
Pine, Aleppo	*Pinus halepensis* Mill.
Pine, Corsican	*Pinus laricio*
Pine, Jack	*Pinus banksiana* Lamb.
Pine, Loblolly	*Pinus taeda* L.
Pine, Lodgepole	*Pinus contorta 'latifolia'*
Pine, Radiata	*Pinus radiata* D. Don
Pine, Red	*Pinus resinosa* Ait.
Pine, Scotch	*Pinus syvestris* L.
Pine, Slash	*Pinus elliotii* Engelm.
Pine, White	*Pinus strobus* L.
Pine, Western yellow	*Pinus ponderosa* Doug.
Pineapple	*Ananas comosus* L.
Pittosporum	*Pittosporum tobira* (Thumb.) Ait.
Plum or Prune	*Prunus domestica* L.
Podocarpus	*Podocarpus macrophyllus* (Thumb.)
Poinsettia	*Euphorbia pulcherima* Willd.
Poplar	*Populus grandidentata* Michx.

Potato, Irish	*Solanum tuberosum* L.
Potato, Sweet	*Ipomoea batatas* (L.) Lam.
Radish	*Raphanus sativus* L.
Raspberry	*Rubus idaeus* L.
Rhododendron	*Rhododendron* sp.
Rice	*Oryza sativa* L.
Rose	*Rosa odorata* (Andr.)
Rye	*Secale cereale* L.
Salvia	*Salvia splendens* Ker-Gawl.
Sansevieria	*Sansevieria laurentii*
Schefflera or Umbrella tree	*Brassaia actinophylla* Endl.
Siratro	*Macroptilium atropurpureum*
Snapdragon	*Antirrhinum majus* L.
Sorghum	*Sorghum vulgare* Pers.
Soybean	*Glycine max* (L.) Merr.
Spathiphyllum	*Spathiphyllum* sp.
Spider plant	*Chlorophytum comosum* (Thumb.)
Spinach	*Spinacia oleracea* L.
Spirea	*Spiraea nipponica* Maxim.
Spruce, Black	*Picea mariana* (Mirp.) B.S.P.
Spruce, Colorado	*Picea pungens* Engelm.
Spruce, Engelmann	*Picea Engelmanii*
Spruce, Norway	*Picea abies* L.
Spruce, Red	*Picea rubens* Sarg.
Spruce, Sitka	*Picea sitchensis* (Bong.)
Spruce, White	*Picea glauca* (Moench)
Squash	*Cucurbita pepo var. melopepo* L. Alef
Statice	*Statice limonium* sp.
Strawberry	*Fragaria* sp.
Stylo	*Stylosanthes humilis*
Sugarcane	*Saccharum officinarum* L.
Sycamore	*Platanus occidentalis*
Syringa	*Syringa* sp.
Taxus	*Taxus x media*
Tea	*Camellia sinensis* L.
Timothy	*Phleum pratense* L.
Tobacco	*Nicotiana tabacum* L.

Tomato	*Lycopersicon esculentum* Mill.
Turnip	*Brassica rapa* L. *var.* Rapifera group
Viburnum	*Viburnum suspensum* Lindl.
Violet, African	*Saintpaulia ionantha* H. Wendl.
Walnut	*Juglans regia* L.
Watercress	*Nasturtium officinale* R. Br.
Watermelon	*Citrullus lanatus* (Thumb.) Matsum. and Nakai
Wheat	*Triticum aestivum* L.
Willow	*Salix* sp.
Yucca	*Yucca elephantipes* Regel

Appendix 4

Useful Conversion Factors and Data

Column A	Column B	To obtain B multiply A by	To obtain A, multiply B by
	Length		
Inches	millimeters	25.4	0.0393
Inches	centimeters	2.54	0.3937
Inches	meters	0.0254	39.37
Feet	centimeters	30.48	0.0328
Feet	meters	0.3048	3.28
Yards	centimeters	91.44	0.01094
Yards	meters	0.9144	1.09361
Miles	meters	1609.3	0.000621
Miles	kilometers	1.6093	0.6214
	Volume		
Cubic centimeters[1]	cups	0.00422	236.967
Cubic centimeters[1]	drops	0.05	20
Cubic centimeters[1]	teaspoons	0.20	4.928
Cubic centimeters[1]	tablespoons	0.0676	14.787
Cubic centimeters[1]	fluid oz.	0.0338	29.574
Cubic centimeters[1]	pints[2]	0.0211	473.176
Cubic centimeters[1]	quarts[2]	0.00106	946.342
Cubic centimeters[1]	gallons[2]	0.00026	3785
Fluid ounces	cubic centimeters	29.574	0.03381
Fluid ounces	teaspoons	6.67	0.1501
Fluid ounces	tablespoons	2	0.5
Fluid ounces	cubic inches	1.805	0.5541

Fluid ounces	liters	0.0296	33.815
Fluid ounces	cups^2	0.1250	8
Fluid ounces	pints^2	0.0625	16
Fluid ounces	quarts^2	0.0313	32
Fluid ounces	gallons^2	0.0078	128
Quarts^2	cups	4	0.25
Quarts^2	fluid ounces	32	0.03125
Quarts^2	pints	2	0.5
Quarts^2	liters	0.9463	1.0567
Quarts^2	cubic centimeters	946.358	0.00106
Quarts^2	cubic inches	57.75	0.01732
Quarts^2	gallons^2	0.25	4
Gallons^2	pints^2	8	0.125
Gallons^2	quarts^2	4	0.25
Gallons^2	Imperial gallons	0.8327	1.2009
Gallons^2	fluid ounces	128	0.0078
Gallons^2	liters	3.785	0.2642
Gallons^2	cubic inches	231	0.00433
Gallons^2	cubic feet	0.1337	7.481
Gallons^2	cubic yards	0.00495	201.974
Cubic feet	liters^2	28.316	0.0353
Cubic feet	quarts^2	29.922	0.0334
Cubic feet	gallons^2	7.481	0.1337
Cubic feet	cubic inches	1728	0.00058
Cubic feet	cubic yards	0.037	27
Cubic feet	cubic meter	0.028	35.315
Cubic yards	cubic feet	27	0.037
Cubic yards	cubic meters	0.7646	1.3079
Cubic yards	quarts^2	807.9	0.00124
Cubic yards	gallons^2	202	0.00495
Cubic meter	cubic feet	35.3147	0.0283
Cubic meter	gallons^2	264.1721	0.0038

Weight

| Grains | grams | 0.0648 | 15.4324 |
| Grains | ounces (avoir dupois) | 0.00229 | 437.5 |

Grams	kilograms	0.001	1000
Grams	ounces	0.0353	28.35
Grams	pounds	0.0022	453.5924
Ounces	pounds	0.0625	16
Ounces	kilograms	0.0283	35.27
Pounds	kilograms	0.4536	2.205
Pounds	ton (short)	0.0005	2000
Pounds	ton (long)	0.000455	2240
Pounds	ton (metric)	0.000453	2205

Water Measurement

Hectare meters	acre inches	97.28	0.0103
Hectare meters	acre feet	8.11	0.123
Acre inches	cubic meters	102.8	0.0973
Gallons[2] per minute	cubic meters per hour	0.227	4.40

Temperature Conversions

Degrees fahrenheit = (degrees centigrade $\times$ 1.8) + 32
Degrees centigrade = (degrees fahrenheit − 32) $\times$ 0.56

Useful Facts of Soil and Water

1 cubic foot of soil in place weighs 70-105 lb.
1 acre slice of mineral soil to plow depth (62\3″)2,000,000 lb
1 cc or milliliter of water weighs 1 gram
1 liter of water weighs 1 kilogram
1 gallon[2] of water weighs 8.345 lb or 3785 gm or 58,417 grains
1 gallon[2] of water occupies 231 cu in or 0.1337 cu ft[2]
1 cu ft of water weighs 62.374 lb and = 7.48 gal[2]
1 cu yd of water weighs 1,685 lb and = 202 gal[2]
1 acre in of water weighs 226,384 lb and = 27,154 gal[2]
1 acre ft of water weighs 2,716,610 lb and = 325,851 gal[2]
1 cu meter of water weighs 2,205 lb and = 264.1721 gal[2]

---

[1]One cubic centimeter = 1 millimeter.
[2]U.S. fluid.

Appendix 5

Sources for Diagnostic Tests and Equipment

Company Name and Address	Supplies
Abbeon Cal Inc. 123 Gray Ave. Santa Barbara, CA 93101	W
Agdia, Inc. 30380 County Road 6 Elkhart, IN 46514	Kv
Ag Leader Technology 1203A Airport Road Ames, IA 50010	MonGP
Agratech 2131 Piedmont Way Pittsburg, CA 94565	Clc
Agri-Diagnostics Assoc. 2611 Branch Pike Cinnaminson, NJ 08077	Kfu
Agrodynamics[1] 10 Alvin Court East Brunswick, NJ 08816	Kfu
A. H. Anderson Co. P.O. Box 1006 Muscogee, OK 74402	MonpH, MonCond
Alltech Assoc. Inc. 2051 Waukegan Rd. Deerfield, IL 60015-1899	An

Aquaterr Instruments
P.O. Box 459
Fremont, CA 94537

Mcond, T, Msm

Arthur H. Thomas Co.[1]
P.O. Box 99
Swedesboro, NJ 08085-0099

Lc, Ls, MpH, Mie, Mcond.
Mi, Hl, An, Ag, R, T, MonT,
Ts, RecT, RecH

Automata
16216 Brooks Rd.
Grass Valley, CA 95945

Df,W, T, MonT, MonIr,
MonSm, MonR, MonIn

Ben Meadows Co.
3589 Broad St.
Atlanta, GA 30341

Hl, MonH, MonpH, MonT,
MonWs, Rg, Ss, T, Wg

Campbell Scientific
815 W. 1800 N.
Logan, UT 84321-1784

W

CEA Instruments
16 Chestnut St.
Emerson, NJ 07630

MonCO2, Ag

CID, Inc.
4018 N.E. 112th Ave.
Vancouver, WA 98682

ACO_2, Ag, Apho

Clements Associates, Inc.
R.R. #1 Box 186
Newton, IA 50208

Ss

Cole-Palmer Instrument Co.
7425 N. Oak Park Ave.
Niles, IL 60714

ACO_2, H, T, M, Mcon, MpH,
RecH, RecR, RecT, Sa, W

Concord, Inc.
2800 7th Ave. N.
Fargo, ND 58102

Ss

Curtin Matheson Scientific[1]
6301 Hazeltine National Dr. No. 100
Orlando, FL 32822-5119

Lc, Ls, MpH, Mcond, Mi,
Hl, R, T, RecT, RecH, Ts

Davis Engineering
8217 Corbin Ave.
Canoga Park, CA 91306

MonT, MonH, MonSR, MonIr

Decagen Devices, Inc.   MonSm
P.O. Box 835
Pullman,WA 99163

DeltaTrak, Inc.   MonT, MonH, MpH,T, H
P.O. Box 398
Pleasanton, CA 94566-0039

EM Science   Ts
P.O. Box 70
Gibbstown, NJ 08027

Environmental Sensors, Inc.   MonSm
13240 Evening Creek Dr. S.
San Diego, CA 92128

Enzytec, Inc.   Ki
415 E. 63rd St., Suite 104
Kansas City, MO 64110

Extech Instruments Corp.   Mcond, MpH, T, H, R, Mlt,
335 Bear Hill Rd.   Mi, Mh
Waltham, MA 02154

Fisher Scientific[1]   Lc, Ls, MpH, Mcond, Mie,
P.O. Box 4829   Hl, An, Ag, Mi, R, Ts,
Norcross, GA 30091   RecH, RecT

Frostproof Growers Supply, Inc.   Ap, Rg, Ss, Msm, It, Hl, W,
512 N. Scenic Highway   Kp, MonT, Wg, As, Mcond,
Frostproof, FL 33843   RecT, RecH, H, HL, Kci, R

Forestry Suppliers, Inc.   Msm, Mi, Te, R, Sw, Si, Ks,
P.O. Box 8397   Gd, Kw, Mi, It, Wm, MpH,
Jackson, MS 39204   Mcond, Wg, RecT, RecH, W

Gemplers   Ag, It, Mcond, RecT, T
P.O. Box 270
Mt. Horeb, WI 53572

Hach Co.   Ks, Kp, Kw, MpH, Mi, Ls,
P.O. Box 389   Lc, Ss
Loveland, CO 80539

Hydro-Gardens, Inc.   Clm, Gd, Hl, Si, Mcond,
P.O. Box 9707   T, As, Aw, IT, Ss
Colorado Springs, CO 80932

Irrometer Co. P.O. Box 2424 Riverside, CA 92516	Te, MonSm
LaMotte Chemical P.O. Box 329 Chestertown, MD 21620	Ks, Ls
Larson Systems, Inc. 103 SE 16th St. Ames, IA 50010-8001	MonGP
Li-Cor P.O. Box 4425 Lincoln, NE 68504	Ag, MonSR, MonT
Marvin D. Kauffman 35178 Balboa PL. S. E. Albany, OR 97321	Ss
Millipore Corp. P.O. Box 9125 Bedford, MA 01730	Kh, Kf
Monitor and Control Resources 22257 NE Seventh St. Redmond, WA 53065	MonT
Myron L. Co. 6231C Yarrow Dr. Carlsbad, CA 92009-4893	MpH, Mcond, Ki
Neogen Corp. 620 Lesher Pl. Lansing, MI 48912-1509	Df, Kw, MpH
Oakfield Apparatus, Inc. P.O. Box 65 Oakfield, WI 53065	Ss
Pest Management Supply, Inc. 311 River Dr., Hadley MA 01035	Df, W, Hl, Rg, It
Precision Moisture Instruments Inc. 100-4243 Glanford Ave. Victoria, B.C., V8Z 4B9, Canada	Msm

Rotronic Instrument Corp.   MonH, MonT
160 E. Main St.
Huntington, NY 11743

Sensor Instruments Co. Inc.   W, Df, MonT, MonH,
41 Terrill Park Dr.   MonR, MonSR
Concord, NH 03301

Sensorgational International   Ss, Mcond, Smo, W, T, H,
P.O. Box 4633   Mlt, MonSm, Aw, MonSr
Hayward, CA 94540

SKC Inc.   Gd, Ag, Sa
RR1 Box 334,
Eighty Four, PA 15330-9614

Skye Instruments Ltd.   Mlt, MonSR, MonSm, MonTe,
Unit 5/6, Ddole Industrial Estate Smo, T, W
Llanddrindod Wells,
Powys LD1 5DF, UK

Soil Measurement Systems   MonTe
7266 N. Oracle Rd., Suite 170
Tucson, AZ 85704

Spectrum Technologies, Inc.   Mch, MpH, Mcond, Mie,
12010 Aero Dr. MNO_3,Mi, MK, MNa,R,
Plainfield, IL 60544   Clm, Rg,Wg, Ss, Spl, Rg
           Wg, W

Teletemp   MpH
P.O. Box 5160
Fullerton, CA 92635

TMI Technical Marketing   Comp
1332 Sadlier Circle, E. Dr.
Indianapolis, IN 46239

Triangle Micro Systems Inc.   Clc
2716 Discovery Drive
Raleigh, NC 27604-1850

Vaisala Inc.   MonCO2, H, T, Wg
100 Commerce Way
Woburn, MA 01801

VWR Scientific[1]   Lc, Ls, MpH, Mie, Mcond,
P.O. Box 669967   Mi, Hl, R, T, MonT, MonH
Marietta, GA 30066

Wadsworth Control Systems   Clc, W
5541 Marshall St.
Arvada, CO 80002

Westgate Agronomics   Comp, Kp, Msm, As, T,
1015 Pitner Ave.   Clc, MpH, Mcond
Evanston, IL 60202

[1]One of several branch locations.

Key to Abbreviations

A = Analyzers
   ACO_2 = carbon dioxide
   Ag = gas analyzers
   Ah = herbicide
   An = nitrogen
   Ap = plant
   Apho = photosynthesis
   As = soil
   Aw = water

Clc = Climate control for greenhouses
Clm = Climate measuring devices
CO_2 = Carbon dioxide measuring devices
Comp = Devices for measuring soil compaction

Df = Disease forecasters

Gd = Gas detectors

H = humidity measuring instruments
Hl = Hand lens

Ir = Irrigation controls
It = Insect scouting equipment

K = Kits
   Kc = chlorine

Kf = fungicide detection
Kfu = fungus detection
Kh = herbicide detection
Ki = insecticide detection
Kp = plant analyses
Ks = soil fertility
Kv = virus detection
Kw = water analyses

L = Laboratory
Lc = chemicals
Ls = supplies

M = Meters
Mch = chlorophyll
Mcond = conductivity
MCO_2 = carbon dioxide
Mie = ion electrode
MK = potassium
Mlt = light
MNa = Sodium
MNO_3 = nitrate–N
MpH = pH
Msm = soil moisture
Mi = Microscopes
Mon = Monitoring equipment
$MonCO_2 = CO_2$
MonCond = conductivity
MonGP = global position
MonH = humidity
MonIr = irrigation
MonpH = pH
MonR = rainfall
MonSm = soil moisture
MonSR = solar radiation
MonT = temperature
MonTe= tensiometer

R= Refractrometers
Rec = recorders
RecH = humidity
RecR = rainfall

    RecSm = soil moisture
    RecT = temperature
    Rg = Rain gauge

S = Sampling equipment
    Sa = air
    Si = insect
    Sm = soil moisture
    Sp = pest
    Ss = soil
    Sw = water
    Smo = soil moisture measuring equipment

T = Temperature measurement
Te = Tensiometers
Ts = Test strips for measuring pH or nutrients
W = Weather stations
Wg = Wind gauge
Wm = Water measuring devices

Index

Page numbers followed by the letter "t" indicate tables; page numbers followed by the letter "f" indicate figures.